Bad Roads and Poor Rations:

Fifty-Nine Wargaming Scenarios for the North American War of 1812

Adrian Mandzy, Ph.D

Bad Roads and Poor Rations: Fifty-Nine Wargaming Scenarios for the North American War of 1812
By Adrian Mandzy, Ph.D
Cover from Adrian Mandzy
This edition published in 2022

Winged Hussar is an imprint of:

Winged Hussar Publishing, LLC
1525 Hulse Rd, Unit 1
Point Pleasant, NJ 08742

Copyright © Winged Hussar Publishing
ISBN 978-1-950423-30-9
LCN 2022932088

Bibliographical References and Index
1. History. 2. North America. 3. War of 1812

Winged Hussar Publishing, LLC All rights reserved
For more information
visit us at www.wingedhussarpublishing.com

Twitter: WingHusPubLLC
Facebook: WINGED HUSSAR Publishing LLC

This book is sold subject to the condition that it shall not, by way of trade or otherwise, be lent, resold, hired out, or otherwise circulated without the publisher's prior consent in any form of binding or cover other than that in which it is published and without a similar condition, including this condition, being imposed on the subsequent purchaser.

The scanning, uploading, and distribution of this book via the Internet or via any other means without the permission of the publisher is illegal and punishable by law. Please purchase only authorized electronic editions, and do not participate in or encourage electronic piracy of copyrighted materials. Your support of the author's and publisher's rights is appreciated. Karma, its everywhere.

Table of Contents

Introduction	1

St. Lawrence/ Eastern Lake Ontario — 3

Gananoque, small skirmish - less than 500 participants	4
St. Regis, small skirmish - less than 500 participants	7
Ogdensburg, skirmish - less than 1,000 participants	12
Cranberry Creek, small skirmish - less than 500 participants	17
Chateauguay, battle - 4,000 participants	21
Crysler's Farm	25
battle - 5,000 participants	26
small skirmish - 450 participants	28
Lacolle Mill, battle - 5,000 participants	31
Plattsburgh	36
large battle - 15,000 participants	37
naval engagement - 30 warships (10 small ships and 20 gunboats)	38
battle - less than 5,000 men	39
small skirmish - less than 500 men	40
sortie on the rocket battery	42

Central Lake Ontario — 46

York, battle - 3,500 participants	47
Sackett's Harbor, battle - 3,500 participants	52
Sodus Point, small skirmish - less than 500 participants	56
Pultneyville, large skirmish - more than 500 participants	61
Oswego, small battle - 1,500 participants	64
Big Sandy, small skirmish - less than 500 participants	67
Kingston, hypothetical attack	70
naval engagement - 17 warships (1 extra large, 2 large, 2 medium, 12 small ships)	70
large battle - more than 5,000 participants	72

Niagara Peninsula/ Western Lake Ontario — 77

HMS *Caledonia* and *Detroit,* small skirmish - less 500 participants	78
Queenston Heights	82
battle - 3,500 participants	82
small skirmish - 60 participants	86
Ft. George, battle - 4,300 participants	88
Stoney Creek, battle - 3,500 participants	93
Beaver Dams	98
small battle - 1,500 participants	99
small skirmish - 3 participants	101
Ft. Schlosser, small skirmish - 70 participants	103
Black Rock, small skirmish - less than 500 participants	105
Burlington Races, naval engagement - 13 warships (4 medium, 9 small ships)	108
Chippawa, battle - 5,000 participants	112
Lundy's Lane, large battle - 6,000 participants	116
Ft. Erie	121
large battle - 6,000 participants	123
large skirmish - 600 to 840 participants (depending on reinforcements)	125

large battle - 7,500 participants	127
Cook's Mill, small battle - 1,650 participants	130

Western Great Lakes — 133

River Raisin, small battle - 2,000 participants	134
Ft. Meigs	138
Battle - 4,500 participants	139
Small Battle - 1,720 participants	141
Large Skirmish - 600 participants	142
Large Skirmish - 850 participants	143
Small Skirmish - 6 participants	144
Put-In-Bay, naval engagement - 14 warships (12 small ships, 2 tiny craft)	146
Thames, large battle - 5,000 participants	150
Mackinac, small battle - 1,000 participants	154

East Coast — 158

Frenchtown, small skirmish - 160 participants	159
Havre De Grace, small skirmish - 400 participants	162
Craney Island, battle – 3,000 participants	165
Bladensburg, large battle - 10,000 participants	168
Caulk's Field, small skirmish - 307 participants	173
North Point	177
large skirmish - 800 participants	178
large battle - 5,800 participants	180
small skirmish - 90 participants	182
Kirby's Windmill, small battle - 1,300 participants	185
Point Peter, small battle - 1,326 participants	187

Gulf Coast — 191

Ft. Bowyer, small skirmish - 90 participants	192
Lake Borgne, naval engagement, 47 warships (47 tiny craft)	195
Villere Plantation, large battle - 5,600 participants	198
New Orleans, large battle - 15,000 participants	202

Appendix #1
Available Figures — 210

Appendix #2
List of Illustrations — 214

Appendix #3
List of Maps — 216

Appendix #4
Converting Pendraken 10mm British Foot to US Infantry — 218

Introduction

The War of 1812, fought between the summer of 1812 and January 1815, took place primary and in North America and pitted the forces of the United States against those of Great Britain. For that reason, the military actions in the Caribbean, or those fought along the frontier, such as Tippecanoe (1811), are not included in this scenario guide.

In terms of numbers of troops employed, the battles were a side-show campaign of the much larger Napoleonic Wars being fought in Europe. European settlements in and around the Great Lakes were still small and logistics remained a nightmare for both sides. As both the British and Americans were ill-prepared to fight the war, each side augmented their standing armies with militias and volunteers. Cavalry was rarely used in its traditional battlefield role and artillery was limited. Fresh water navies grew overnight on the Great Lakes and as the war progressed, the British and then the United States employed Native Americans.

The War of 1812 had a number of features which make it well suited for wargaming. The armies were small. Thus, the gamer can easily recreate even the largest battles of war, which were no bigger than division sized actions fought in Europe, on a 6 by 4 foot table. It is also a colorful period. Marines and naval raids, rockets, pirates, and extravagant multi-colored uniforms adds to the visible spectacle of a War of 1812 game. British line and some of the better equipped militia were almost indistinguishable from the Crown forces that fought in Spain and Waterloo, so those with already painted armies can use them on both sides of the Atlantic. Similarly, militia troops from the American Revolution, the Alamo, or even the French and Indian Wars could be used if one avoids too many tricorn hats. The variety of troops and the militia on both sides can lead to a more challenging game, and also serves to reduce the monotony of painting. The military leadership was varied as well, as both sides fielded brilliant officers, as well as incompetents. Thus, the stage was set for amateurs and professionals, along with fools and heroes, to battle for three years.

Though various scenarios for the War of 1812 book have appeared in print over the last thirty years, including at least two now long out of print scenario/wargaming guides to the War of 1812, they generally tend to focus on the larger battles. Unfortunately, this provides a distorted view of the war and ignores most of the fighting that occurred. In recent years, rules tend to look at smaller size actions and skirmish gaming has grown exponentially. Large flow charts and simulations have, for the most part, been replaced with games that can be completed in a few hours. Games remain serious, but the focus is on fun.

As control of waterways was the defining factor of the conduct of the war, the three major theaters of operations were: 1) the Great Lakes, especially Lake Ontario, 2) the East Coast, including the Chesapeake, and 3) the Gulf Coast. Most of the scenarios presented in this volume are for recreating land-based actions, but we included naval fights along the fresh waterways as they were a critical component to land based operations. For completion, a few pure naval actions along the Great Lakes and the Chesapeake were also included in this work.

The scenarios presented in this work are organized by battles and theaters of operations. By organizing the book this way, gamers who build particular units can quickly use the same figures in different scenarios. The other advantage of this system is that the historical fights, especially the larger ones, often provide the background for more than one type of tabletop recreation. Rather than repeat the historical background for each scenario, we include it once and then make suggestions on how best to recreate various parts of the battle. By including a number of tabletop recreations, you as a gamer have options. If you are just getting into the period, you can still run a game with a handful of figures and have a good time. Then, as your interest in the period grows, you add more figures. As you complete each particular batch of miniatures, you can integrate them into your force and game the next larger scenarios. In time, you will have painted so many figures that you cover the table and be able to game the largest battles of the war.

The scenarios are designed to run with any rule set and were not written to fit with any particular game system. While the author uses a variety of rule sets to recreate the land actions, including Ken Cliffe's now defunct *All The King's Men* rule set, as well as his own home-grown variants, Warlord Games new *Black Seas* rules provides a great way to replicate ship to ship or ship to shore combat.

The actual historical orders of battles have been drawn from existing sources, but often these sources conflict with each other. The scholarship on the War of 1812 is varied and while some battles have multiple monographs, other actions are poorly documented. We have used our best judgement and sought to provide numbers that are both supported by the existing literature and which would make for a balanced game. Those troops and numbers appearing in italics represent our best guess estimates.

This wargame scenario guidebook came about with my long-term interest in wargaming and the War of 1812. Growing up in western New York twenty minutes from Lake Ontario, in my youth I spent many hours exploring places like Ft. Ontario, Pultneyville, and Ft. George. I was fascinated by the relics of the war, like the bayonet an older friend found at Sackett's Harbor or the part of a musket barrel recovered from Carleton Island. Graduate studies in archaeology and history took to me first to Michigan and then later Toronto, were I studied and wrote about Central Europe. Along the way I painted the 25mm Wargames Foundry War of 1812 figures and gamed in people's basements and conventions like SpartaCon and HistoriCon. For the last twenty years I have worked in the field of battlefield archaeology in both the US and Europe, replaced my older figures with the far more complete 40mm Sash and Saber range of figures, and had the pleasure of wargaming a historical miniatures battle on the 1813 Leipzig Battlefield with my project co-director.

I would like to publicly thank a number of people who helped with this wargaming scenario guide. Richard Barbuto, who has numerous works on the 1813 and 1814 Niagara campaign is among the most insightful military historians of our generation, was kind enough to review both the concept and work itself and made a number of suggestions. Paul Lear, the site director of Ft. Ontario, was a treasure trove of information and provided access to maps, photographs, and sources. Ralph Naveaux, whose has written on the Battle of the Raisin, provided maps, photographs, and illustrations. Richard Colton and Lawrence Babits shared their experience on all things technical, especially artillery. I also want to thank Craig Charron, a gamer and War of 1812 reenactor I first met in my first semester of graduate school and Karl Shanstrom, another fellow gamer which helped to create the work you now hold in your hand. Chris Hughes, the owner and sculptor of Sash and Saber Casting, provided many of the 40mm figures pictured in the book and Forrest Harris of Knuckleduster Miniatures provided photographs of his 28mm line of figures. Alan Perry of Perry Miniatures took some photographs of their extensive British Napoleonic range and share them with me. Ken Osen of W Britain shared photographs of their 54mm War of 1812 figures. Pendraken Miniatures generously provided samples of their 10mm Napoleonic British figures, some of which were converted to portray US Infantry. Other manufactures, such as Wargames Foundry, Britannia Miniatures, and GHQ Models, also shared pictures of their miniatures. I also want to thank Warlord Games for their donation of figures and rules, including the new *Black Seas* system. Many of their photographs appear in this scenarios book. Other wargamers, such as Dave Abraham, Roger Chrysler, Tod Kershner, Chris Ngiau, Harold Oney, Mike Vasile, and Richard Watts took pictures of their figures and shared them with me.

I also need to thank Christian Wright, my undergraduate research fellow, who read the text and made a number of suggestions. During the course of this project, Christian learned basic graphic design and drew most of the maps that appear in this volume. My editor, Vincent Rospond, is not only one of the nicest people that I know, but after a number of books and articles, he is by far the most pleasant editor that I have worked with over the last thirty years. It also goes without saying that any and all errors within this volume are solely the fault of the author. Finally, I also want to thank my wife Natalia and children Athena and Theo who gave me the time to write and suffered as I dragged them to War of 1812 forts, museums, and battlefields, include the closed Ft. Norfolk, which we all got to see through a locked gate.

St. Lawrence/ Eastern Lake Ontario

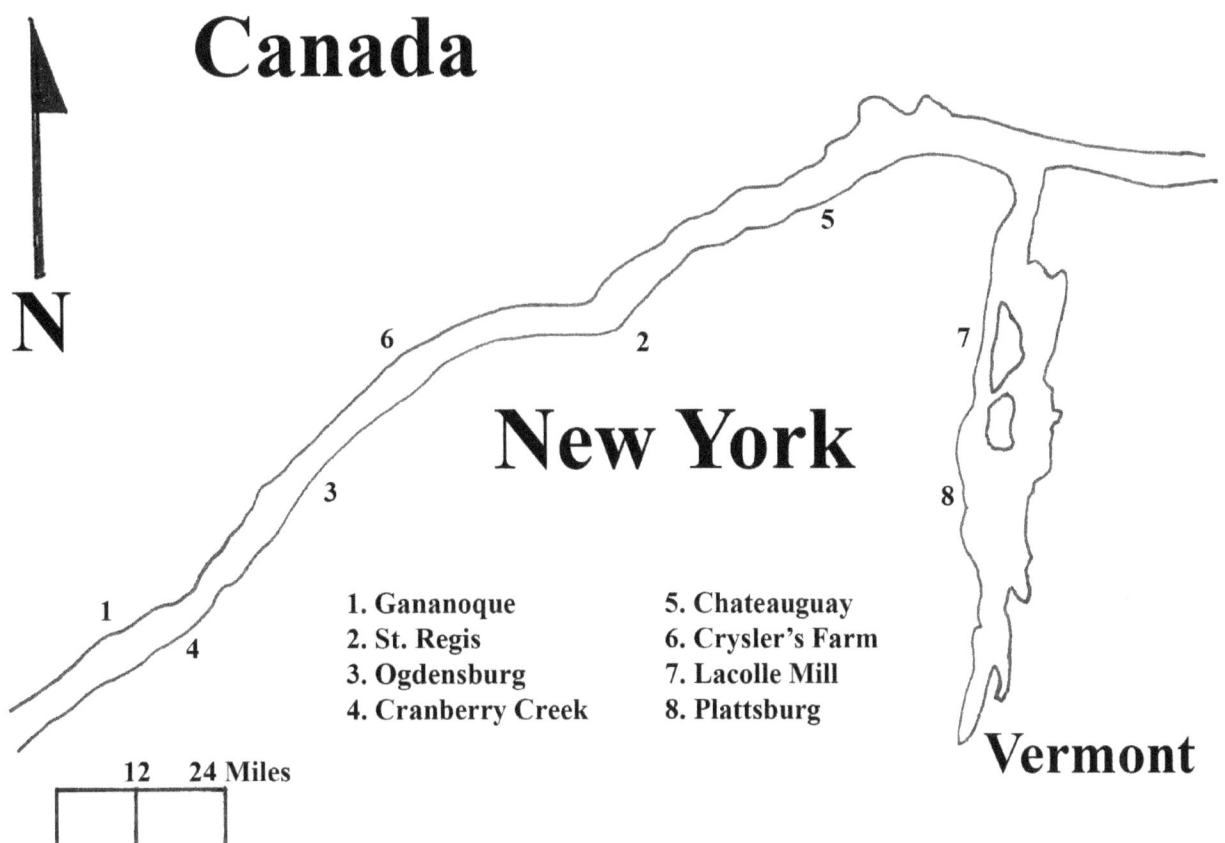

Map 1

Bad Roads and Poor Rations
Raid on Gananoque, 21 September 1812

1. British Light Dragoons - courtesy of *Perry Miniatures*

Overview
Small Skirmish - less than 500 participants

US riflemen and American militia conduct a raid on the Canadian settlement of Gananoque.

Introduction
With the start of the war, Capt. Benjamin Forsyth of North Carolina was sent with a company of men to garrison Sackett's Harbor. Brig. Gen. Jacob Brown, commanding New York State Militia troops, authorized Forsyth to raid the Canadian village of Gananoque. Located along the supply route between Kingston and Montreal, Gananoque grew up along both sides of a stream. In 1812, a stout stone bridge and mill dominated the settlement.

Leaving Sackett's Harbor, the Americans spent the next three days sailing against the headwinds and avoiding patrolling British gunboats. Landing at Sheriff's Point, the American landing was spotted by two Leeds Mounted Militia Dragoons. A shot by a rifleman silenced one of the scouts, while the other was able to ride off and successfully raise an alarm.

The Battle
As the riflemen began their advance on King's Road to the town, the red coated Leeds militia appeared to greet them. The militia lined up and discharged a volley, which killed a rifleman outright and wounded "a few more". Before the militia were able to fire again, the riflemen charged the red coats. Rather than face the charge, the militia unit disintegrated, and the men fled across the stone bridge. A few militiamen were killed as they ran, and the Americans occupied the village. Mindful that British troops from nearby Kingston would soon be dispatched to counter the raiders, Forsyth quickly collected a few prisoners, sixty stands of arms, a few barrels of flints and powder, and set fire to those military provisions they could not carry away with them. Less than an hour since the raid began, Forsyth's men were back on their boats.

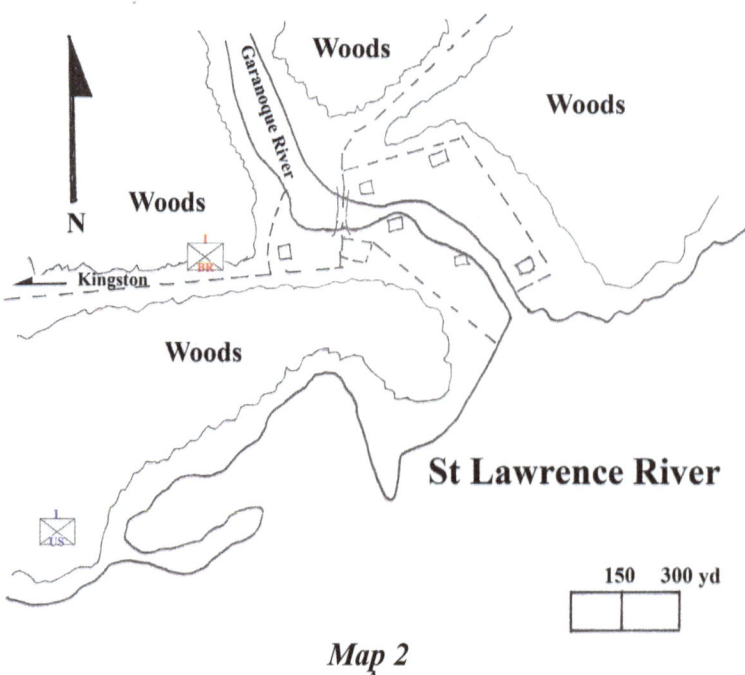

Map 2

Order of Battle[1]

British

Commanding Officer: Col. Joel Stone
2nd Regt. Leeds Militia, 60 men

Reinforcements from Kingston
Canadian Light Dragoons – 1 troop, 40 men
British Line – 2 companies, 120 men
Canadian militia – 2 companies, 120 men

US

Commanding Officer: Capt. Benjamin Forsyth
First US Rifles, 70 men
New York State Militia, 34 men (stayed with the boats)

2. US Rifle Regiment courtesy of Richard Watts

Game Notes

The Raid on Gananoque makes for an interesting tabletop encounter. Although the riflemen individually are known for their marksmanship, both sides are untried in combat. Forsyth was a natural leader and born light infantryman. Stone, a loyal British subject who fled to Canada after the American Revolution, was an elder administrator who despised the southerners and did not distinguish himself on the battlefield. For game purposes, Forsyth should be rated as bold but untried (a D6 die roll could be used to determine how effective he is on the game table – a 1, 2, or 3 slightly above average, while a 4, 5, or 6 would be outstanding). For Stone, a similar die roll would determine if he were below average or average.

---

[1]Troops and troop numbers that are italicized represent the author's best estimate. If a unit's name appears in italics, there is some question if they were present or participated in the engagement.

In our game, we place three wooden houses and a stone house on the south side of the stream. The stone mill and three more houses are placed on the north side of the stream. Six "spoils of war" tokens are placed by the British player in the village – one in the mill, one in the stone house, two in the three houses south of the stream and two in the three wooden houses north of the stream.

To duplicate the riflemen's pillaging of the village, the American player needs to enter a house and spend the turn searching the building to claim a "spoils of war" token. If the American player wants to set fire to a house, they should spend a turn lighting it. No die roll is needed to burn a wooden structure but setting fire to either the stone mill or stone house should require a die roll.

To place additional pressure on the Americans, we limit the game to twelve turns and add British reinforcements. At the start of Turn 7, a D6 die roll is made to see if additional British troops appear on the table. On a roll of a 6, the British show up on the northeast section of the table at the end of the turn in a march column. If no British reinforcements appear, at the start of Turn 8, a new die roll is made; this time a 5 or 6 is needed. This continues for each following turn, with the number needed on the die roll decreasing by one until the British show up automatically on turn twelve.

For Forsyth and his men to claim victory, they need to defeat the militia and return with three "spoils of war" tokens to their boats. For every two buildings the Americans set on fire, they can claim one "spoils of war" token. The British need to keep the Americans from achieving their goals.

Further Reading

Fredricksen, John D. *Green Coats and Glory: The United States Regiment of Riflemen, 1808-1821.* (Old Fort Niagara Association. Youngstown, New York. 2000)

Lossing, Benson J., *Pictorial Field-Book of the War of 1812.* (Harper & Brothers Publishing, New York. 1885)

3. Incorporated Militia, courtesy of Knuckleduster Miniatures

St Regis, 23 October 1812

Overview
Small Skirmish - less than 500 participants

American militia raid the Native American village at St. Regis.

Introduction
Though the 1783 Treaty of Paris clearly established a border between the former colonists and Great Britain, for the most part Native Americans controlled the territory north of the 1768 treaty line. British officials on the frontier sought to preserve the Montreal-centered fur trade and use the Native Americans as a barrier against

4. A View of the Old Church in St. Regis, (Lossing)

the Americans. For the Native Americans, the new border was an abomination and those that resided on both sides of the border hoped to be able to play one side off against the other, much like they had done during the 17th and 18th centuries.

Catholic Mohawks from south of Montreal had founded the settlement of Akwesasne in the mid 18th century. The border set by the 1783 Treaty did not take into account this settlement, and as a result, its residents found themselves on both sides of the international boundary. With the start of the War, the Americans and the British sought to bring the residents of Akwesasne to their side. Though officially neutral, by early October both sides sent military forces into the strategic area.

On 1 October, New York State Militia built and garrisoned a blockhouse at French Mills, which was part of Akwesasne territory. The British, hoping to recruit warriors to their cause, sent troops and agents to St. Regis Point, the principal settlement of the Akwesasne territory. Fearful of Native Americans siding with the British, New York State Militia took it upon themselves to prevent this alliance from occurring.

The Battle
Maj. Guilford Dudley Young and a small detachment of militia from Troy, New York, were led by William Gray, a Native American interpreter, along an unfrequented path to St. Regis. Arriving on the other side of the river from the settlement, the New York State Militia were forced to turn back, as they had no craft with which to cross the river. Undeterred, Maj. Young's force acquired three craft, crossed the river at night, and marched on St. Regis.

Arriving at the settlement, Young split his command into three columns. The militia then surrounded the settlement and waited for dawn to launch their attack. The British had posted pickets, but these failed to see the approaching militia. Both the Native Americans and Crown's soldiers were caught off guard and after a few Canadians were killed by American volleys, the remaining troops surrendered. The Americans proceeded to loot the village and collected numerous spoils, including forty stands of arms, 800 blankets found at the Indian Agent's house, and a British Union Jack. The Americans then withdrew to French Mills.

Bad Roads and Poor Rations

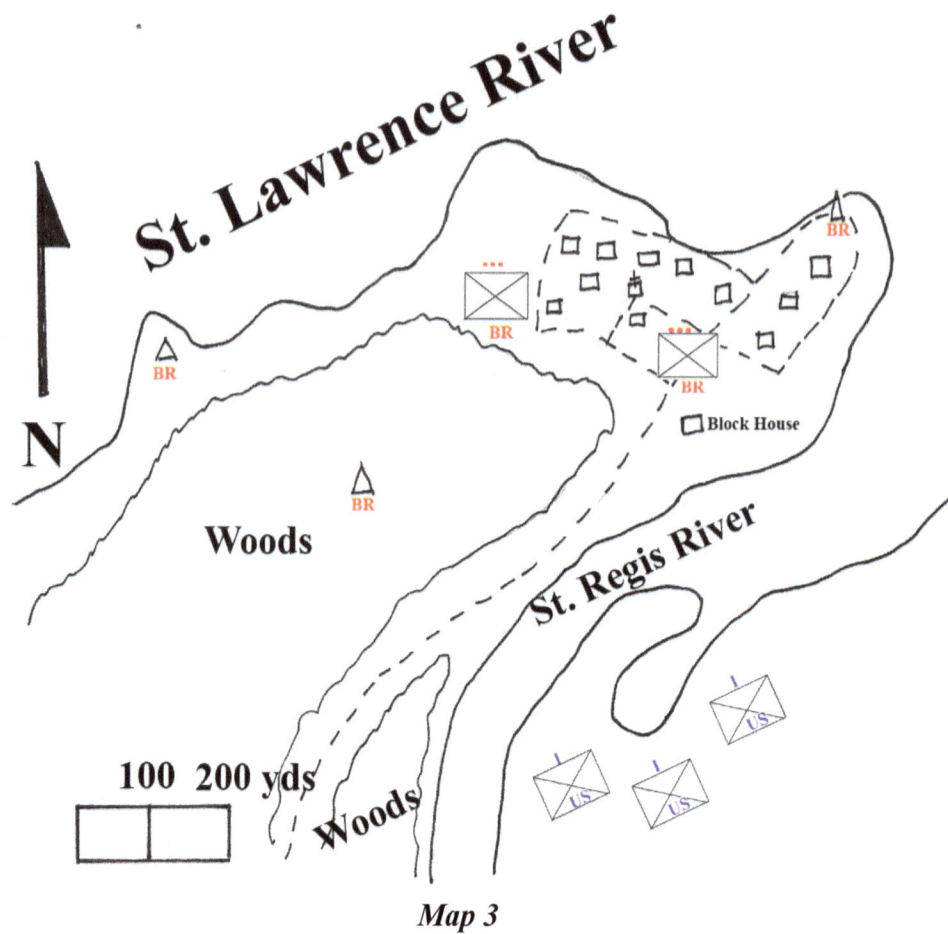

Map 3

Order of Battle

British, 42 men
Commanding Officer: Capt. John McDonell
Company of Canadian Voyageurs – 40 men, disbursed into two groups
de Montigny (resident British agent)

US, 160 men
Commanding Officer: Maj. Guilford Dudley Young
1st Column
    Capt. Lyon
    1 company, *40 men*

2nd Column
    Capt. Tilden
    1 company, *40 men*

3rd Column
    Maj. Young
    Capt. Higbie
    1 company – *40 men*

    Capt. McNeil, 1 company – *40 men*

5. Native Americans, courtesy of Craig Charron

Game Notes

The dawn attack on the settlement of St. Regis is a natural draw to gamers who recreate the war on the frontier. It has Native Americans, a bold dawn raid, three columns, and the possibility of loot. The Canadian troops wore civilian clothing, while it is unclear what the American militia were wearing. For our game, we have officers wearing red and blue coats, but the rest of the troops wear civilian dress. The scenario is designed for ten turns.

In our game, we attempt to duplicate the American movements by a series of die rolls to see if they reach their target at the appointed time using a D6. For each of the three columns, a roll of a 1 or a 2 signifies the particular column is delayed by 1 or 2 turns accordingly. This replicates the dangers of moving militia at night through unfamiliar territory and also gives the British player some hope of beating off the initial attack.

To keep the game moving, the scenario begins with the British troops disordered and scattered in different locations in the village. Both sides can take actions on the first turn. Refer to your favorite rule set on how to recover disordered units. Historically, the American militia units were brittle, in that they can easily break and run if fired upon or charged. The Canadian volunteers in British service were not professional troops and if fired upon from two sides, will automatically surrender.

If the American player is able to capture the village and dislodge the Canadian troops, which happens in two out of three games, their victory is not yet assured. Americans were well known for their fondness in collecting the spoils of war and need to capture a British flag. To duplicate the search for the Union Jack, the Americans need to search each structure.

The American attack was directed against the northern part of the village, where the British had billeted their troops. By the beginning of the 19th century, Native Americans had, for the most part, abandoned their traditional long houses and were building single dwelling houses. Though a wooden church stood in the center of the village, it played no part in the battle. The British had also built a wooden blockhouse in the village, but it also does not appear to have been used during the battle. For our game, we use one wooden blockhouse, one church, two longhouses and eight single story buildings. If players do not have the required three dimensional buildings, simple two-dimensional representations may be easily created from paper or cardboard.

To secure their victory, Americans historically looted the buildings in search of war trophies. After the battle, much political hoopla was made of "the capture" of a British Union Jack which was found at St. Regis. Personal property was also not respected and one 13 year old girl's box was plundered for $3 worth of pennies. In this game, the British player secretly places the Union Jack in a building of their choice before the game begins.

The Native Americans were officially neutral at this stage of the war but could easily be drawn into the fighting, especially when the Americans begin looting their homes. Historically, Native Americans from Akwesasne took up arms against the Americans in the retaliatory attack on French Mills and fought all along the border for the duration of the war on the side of the British.

In our games, we duplicate Native American involvement by rolling a D6 for each house that is looted. On a roll of a 6 the Americans find loot (either one victory point or the Union Jack). On a roll of a D1, however, the Native Americans in that house attack the looting militia. To determine the number of defenders in each structure, a single D6 is rolled for each single-story building, two D6 are rolled for each of the two longhouses, and three D6 are rolled when the Americans loot the church. No dice are rolled when the Americans loot the blockhouse.

Victory Conditions

To determine the victor, the American needs to outnumber the British player in victory points. The following victory points are counted for the Americans:

Each piece of captured loot – 1 point

Captured Union Jack – 5 points

Each British casualty –1 point

A similar point system is used for the British:

Each destroyed American column (either routed/captured/destroyed) – 10 points

Each American casualty – 1 point

6. New State Militia, courtesy of Richard Watts

With such a point system, the American player can still lose the game even if they destroy all of the Canadian troops on the table.

Further Reading
Benn, Carl, *The Iroquois in the War of 1812.* (University of Toronto Press. Toronto. 1998)

Fredricksen, John D., *Green Coats and Glory: The United States Regiment of Riflemen, 1808-1821.* (Old Fort Niagara Association. Youngstown, New York. 2000)

Lossing, Benson J. *Pictorial Field-Book of the War of 1812.* (Harper& Brothers Publishing. New York. 1869)

Ogdensburg, 22 February 1813

Overview
Skirmish - less than 1,000 participants

British raid on the American village of Ogdensburg.

Introduction

Located on the American side of the border, across from the Canadian village of Prescott, the village of Ogdensburg is situated where the Oswegatchie flows into the St. Lawrence River. At the start of the 19th century, a brisk trade developed between these two border towns and continued, for the most part, unabated with the start of the war. The strategic importance of Ogdensburg did not go unnoticed and in early October, Col. Robert Lethbridge, the British commander at Prescott, launched an assault on Ogdensburg, which was met by cannon fire. Before the British came within a quarter mile of the town, they had suffered casualties and withdrew back across the St. Lawrence River.

On 12 October, a company of US Rifles under Maj. Benjamin Forsyth (who was a captain at Granaoque) were posted to Ogdensburg. Forsyth was already making his reputation for bold actions, such as his recent raid on Gananoque, and by placing his command in a place well suited to threaten the British supply route from Montreal to Kingston. From this location

7. Red George Mac Donnell, courtesy of John R. Lee

he could launch further raids against the enemy. The presence of US troops would also help secure the border. The ruins of an old French fort, abandoned since the end of the French and Indian War, were still visible in Ogdensburg, but were ill-suited for the current conflict. Thus, in addition to stationing troops in the area, the Americans decided to construct a new fort, which they named Fort Oswegatchie.

Over the next few months, Ogdensburg served as a base of operations to launch hit and run on the British supply lines. The British conducted similar raids against the Americans. Annoyed by the American military raids originating from Ogdensburg, Lt. Gen. George Prevost, the commander in chief of the British in North America, when visiting Prescott, authorized Maj. George Macdonell to launch a "demonstration" against Ogdensburg. The purpose of the raid was to divert attention from the presence of Prevost in the area, as his presence made a tempting target for the likes of the US Rifles.

The Battle

In the winter, the St. Lawrence had frozen and ice was thick enough for troops to march across. On the morning of the 22nd of February, Maj. Forsyth had watched the British troops drill on the ice, as they had many times before, but as the British formed two columns and began marching across the ice, the Americans began for rally to meet the assault. Progress across the ice was slow and the American defenses began to fire on the right column under Capt. Jenkins. The assault wavered and the Right Column withdrew. At the same time, the larger Left Column's march into town without attracting any resistance. The American Capt. Conkey promptly surrendered his men and his unfired 12-pdr. cannon, which was perfectly positioned to hamper the British assault.

Having successfully entered Ogdensburg, the British received fire from two small American cannons. The first cannon was quickly put out of action when its elevator screw broke, which rendered the gun inoperable, and its crew quickly crossed the Oswegatchie River to join Forsyth. The second gun continued to fire until two of the crew were killed and the rest of the crew was captured.

Having secured the town, the British turned their attention on the American troops within the confines of the ruined French Fort. The British, under a flag of truce, sent a message to Maj. Forsyth demanding his surrender. To which Forsyth replied, "there must be fighting done first". As the British and Americans began to exchange fire, a second British force prepared to storm the fort from the rear. Recognizing that the outnumbered Americans were about to be overwhelmed, Forsyth withdrew his force first to Thurber's Tavern on Black Lake, and then eventually to Sackett's Harbor.

With Ogdensburg securely in British hands, residents from Prescott, along with Native Americans, looted the town. As most of the American civilians had abandoned the settlement, the Canadians helped themselves to whatever private property they could find, while the public property was taken by the British troops. Two armed schooners and two gunboats trapped in the ice were burned and the US barracks were laid to ash. Their goal of neutralizing Ogdensburg achieved - the British withdrew to Prescott. Shortly thereafter, the American civilians returned to their devastated homes while the US Army never did.

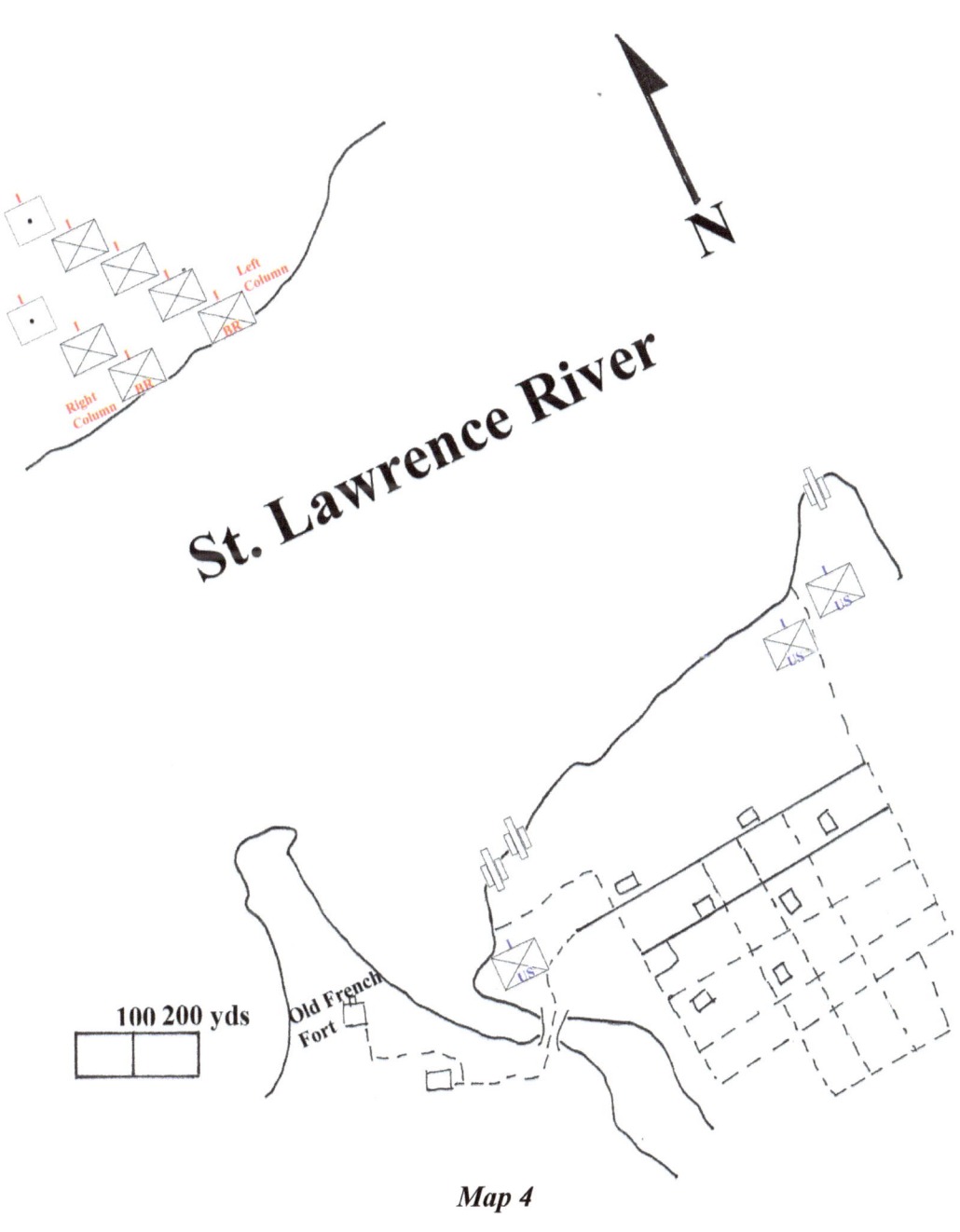

Map 4

Order of Battle

British - 550 men
Commanding Officer: Lt. Col. "Red George" MacDonnell

Right Column, 130 men
Capt. John Jenkins
Glengarry Light Infantry Fencibles, one company, *60 men*
Militia, 70 men
Artillery mounted on sleds, 2 small guns

Left Column, 420 men
Lt. Col. "Red George" MacDonnell
8th Regiment of Foot, 120 men
Royal Newfoundland Regiment, 40 men
Militia, 200 men
Native Americans, 60 men appear on the field only after the British have crossed the river
Artillery mounted on sleds, 2 small guns

US
Commanding Officer: Maj. Benjamin Forsyth

At the Old French Fort
Maj. Forsyth
First US Rifle Regiment, one company, *60 men*
Two 6-pdr. guns, *20 men*

In the Town
Capt. Conkey
New York State Militia – 4 detachments, *180 men*
Artillery – one 12-pdr, two 6-pdrs., *20 men*

8. British in overcoats, courtesy of Roger Chrysler

9. US Militia artillery, courtesy of Richard Watts

Game Notes

This winter raid across the ice makes for a pleasing wargame. The forces are equally matched, and the variety of troop types make it challenging for both the British and American players. The overwhelming majority of troops are militia, which are unpredictable at best, but may let off a decisive volley that changes the outcome of the game.

The terrain is easy to replicate, as a white game mat or sheet makes for a good representation of both the ice and snow-covered landmass. In our games, we use a series of rocks to mark the limits of both the St. Lawrence and the Oswegatchie Rivers. The American troops were never issued overcoats along the northern frontier, so when one fields miniatures in summer uniforms they are correct for this engagement. Canadian troops, on the other hand, did have watch coats, but rather than build a second army in them, we depict them in their red coats. The red coats maybe less accurate, but it does illustrate the early 19th century values of dressing up before battle. Red coats are more impressive on the tabletop, as they were in real life.

At the time of the battle, the ice was too thick for the British cannon to break through, so there is no need to roll for movement over ice. Though the American troops fired their smaller cannon at the British while they were still crossing the St. Lawrence, there is no evidence that it broke the ice to the detriment of the Canadians. If one wishes to recreate a scene similar to that of Alexander Nevsky's Battle on the Ice, where the break in the ice causes the heavily armored Teutonic knights to fall to their deaths in the ice water, it is possible that the energy of the 12-pdr. gun (which was never discharged in the battle) could cause such chaos.

When starting the battle, place the two British columns three movement turns away from the defenders at the start of the game. On each turn, we roll a D6 to see if the American militia activate (a roll of a 5 or 6 being needed). Once the militia activate, they may fire the cannon on the following turn. The poorly trained militia will need to spend two turns loading the 12-pdr. gun. Even one 12-pdr. cannon can be devastating at close range and if the British are still crossing the river, there is a 33% chance that solid shot could break the ice (1 or 2 on a D6). If that happens, then any targeted unit or group of figures would be swept under the River. Once the British player reaches land, this becomes irrelevant.

After the British captured Ogdensburg, the Americans began to fall back. Pressed by overwhelming numbers, historically the riflemen focused on extricating themselves as best as possible. If the game follows this historical pattern and the American player finds themselves in danger of being cut off, they should try to get off the western table edge.

The British focus of the raid was to end Maj. Forsyth's activities and thus safeguard the shipment of British goods down the St. Lawrence. For this reason, the British player should focus on defeating Forsyth, either by capture or by killing him. For each attempt at capture, especially if followed by a series of brave words, the British player gains a point of honor. Then, if the good major is killed, honor is satisfied, and the British can obtain a glorious victory.

Ideally, the American player should hold the town and they have a good chance of meeting this objective. Failing to hold the town, the Americans can still deny the British the satisfaction of killing or capturing Forsyth. The American player can gain victory points if they inflict casualties on the British force. Maj. George Macdonell was under strict orders and if the British suffer significant losses, the repercussions from the attack would be damaging to his career. Thus, if at any time during the game the British suffer more than 20% casualties, or Maj. Macdonell is killed, they need to begin an immediate withdrawal. The player controlling should be encouraged to play it as such with a fighting withdrawal.

Victory Conditions

If the British capture and are able hold the town before suffering 20% casualties, they gain a major victory. The British also can gain a major victory if they kill Forsyth before suffering 20% casualties. The American win a minor victory if the British take the town, but Forsyth is killed. If Forsyth survives and the British are driven off, the Americans gain a major victory. The game plays for 12 turns.

Further Reading

Fredricksen, John D. *Green Coats and Glory: The United States Regiment of Riflemen, 1808-1821.* (Old Fort Niagara Association. Youngstown, New York. 2000)

Hough, Franklin B. *History of Jefferson County.* (Joel Munsell. Albany, NY. 1854)

Lossing, Benson J. *Pictorial Field-Book of the War of 1812.* (Harper& Brothers Publishing. New York. 1869)

Quimby, Robert S. *The United States Army in the War of 1812.* (Michigan State University Press. Lansing, MI. 1997)

Cranberry Creek, 19 July 1813

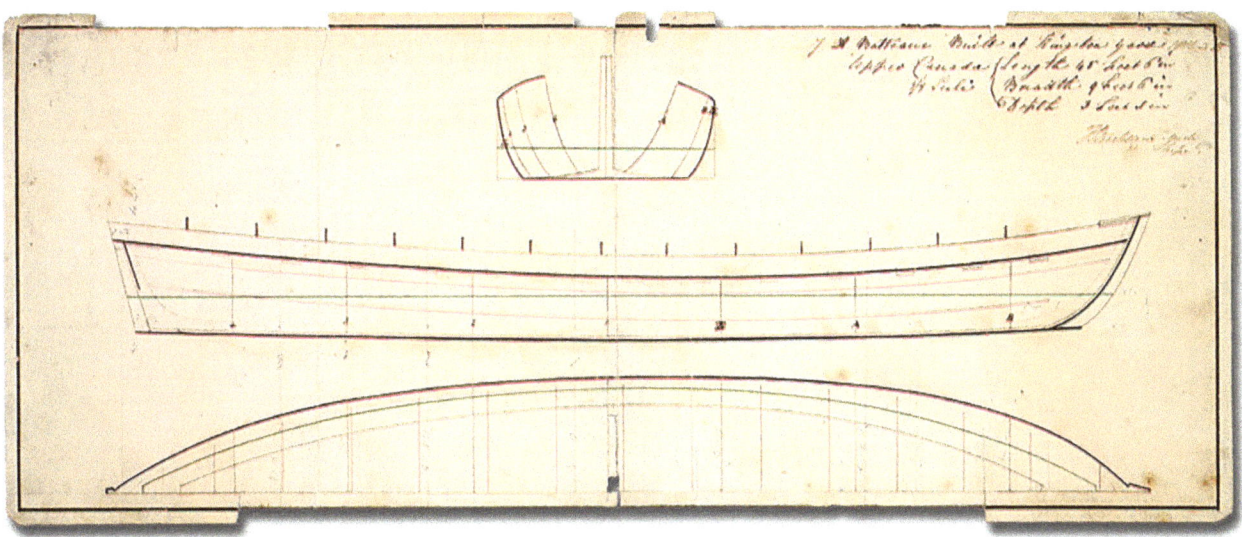

10. 1814 Drawing of a Bateaux, Mrs. Edward Kemp Collection F360 OS 1-8 Archives of Ontario AO5985

Overview

Small Skirmish - less than 500 participants

After US troops captured a British supply convoy, British troops chased the US forces up a creek but failed to recapture their goods.

Introduction

During the War of 1812, the British forces on Lake Ontario were supplied both overland and by boats via the St. Lawrence River. With the support of the American military commanders, local businessman Marius Gilbert obtained formal Letters of Marque and outfitted two boats of privateers on the lake. In early July 1813, the American forces at Sackets Harbor (NY) learned that a shipment of 15 *bateaux* (flat bottom cargo boats) was on the way to Kingston, the major British supply base on the lake. The British convoy was escorted by one gunboat, the *Spitfire,* and was manned by seventy British troops. The two American armed boats surprised the British while they were resting at Simmond's landing and the entire convoy was captured without a shot being fired. The force then retired to Cranberry Creek in New York to divide the spoils. The British prisoners were then marched overland to the American base at Sackets Harbor.

In anticipation of a British counterattack, the American commander pushed his flotilla up Cranberry Creek and began preparations. Using the more than 270 barrels of pork and flour captured in the raid, the Americans proceeded to build a fort. This "pork barrel fort" was strengthen by logs. To prevent the British boats from moving further upstream, tree branches were placed in a bend in the river.

The Battle

Learning of the captured boats, the British sent the 18-gun *Elmira* and several boats to intercept the Americans. Though the water was too shallow for the *Elmira* to push up into Cranberry Creek, 4 gunboats and 2 *bateaux* advanced up the waterway. Upon pressing their advance, the British forces found that the winding creek was barricaded with cut down trees. When the British commander sent troops into the knee-deep water to remove the barricades, the Americans began to fire. The Americans had prepositioned their two gunboats so they could fire on the advancing British, while the forces of the Crown could not bring their forward-facing guns to bear on the American craft due to a bend in the creek. Moreover, by being lower in the water, the British were unable to successfully elevate their heavy cannons while the Americans continued to pelt the British from two positions with small arms and cannon. British forces tried to land and advance along the south bank but were unable to drive the Americans from their log entrenchments and "pork barrel fort". After suffering additional casualties, British abandoned one of their *bateaux* and successfully withdrew from the ambush. After a few days, the Americans eventually remove the captured supplies to Sackets Harbor.

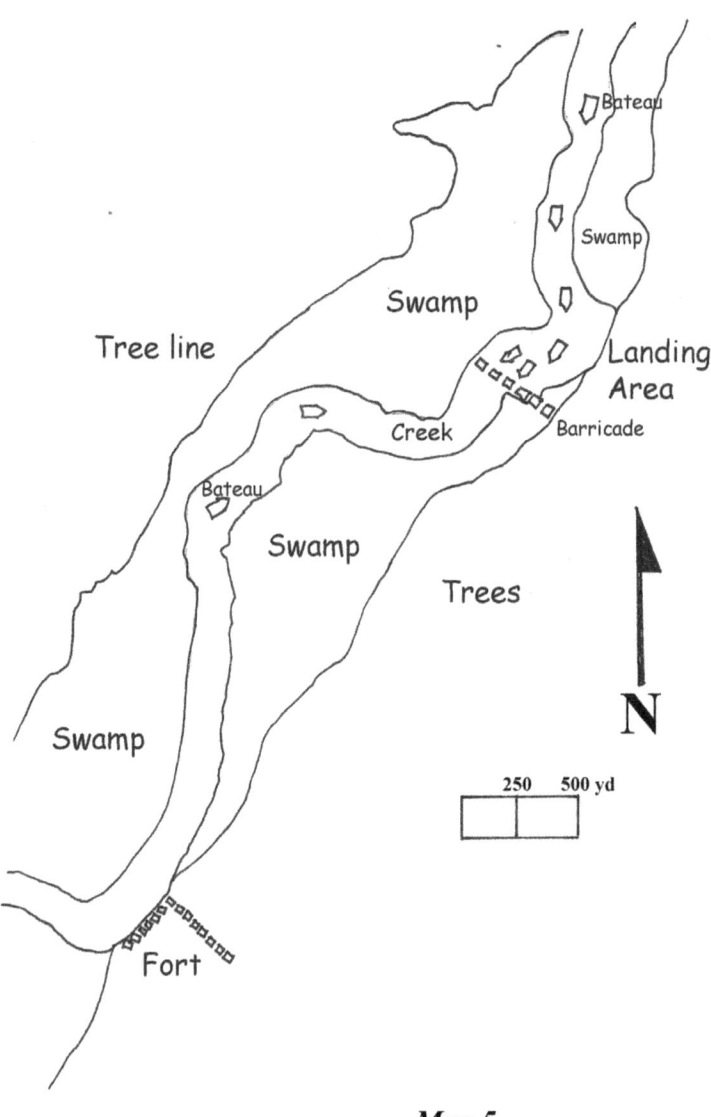

Map 5

Order of Battle

British - 250 men
Commanding Officer: Maj. Frend, 41st Regiment of Foot
Henry O. B. Milnes, Aide-de-Camp to George Prevost

100th Regiment of Foot, detachment, 80 men
41st Regiment of Foot, 2nd Batt., *1 company,* 60 men
8th Regiment of Foot, 1st bn, *1 company,* 60 men
Detachment of Royal Marines, 30 men
Detachment of seamen, 20 men

4 gunboats, 2 *bateaux* pursue the Americans up the creek

US - 120 men

Commanding Officers: Capt. Jehiel Dimmock
Marius Gilbert, privateer

New York State Militia, Capt. John Hoover, 55 men
Twenty-First US Infantry: 21 men from the regiment took part in the raid into Canada.
    Following their return to the US, 15 men and Lt. Burbank marched the Crown prisoners to Sacketts Harbor. Five men from the regiment remained to guard the captured supplies.
First US Rifle Regiment on the Fox under Lt. Hawkins and Sgt. James, 20 men
Twenty-Fifth US Infantry, Lt. Read, 20 men
Ninth US Infantry, Lt. David Perry, 20 men

2 gunboats/privateers, 1 "pork barrel fort"
    Neptune (privateer) - one 6-pdr. cannon, 1 swivel
    Fox - one 6-pdr. cannon, 1 swivel
    Small cannon (1.5 swivel gun) at the "pork barrel fort"

11. New York Militia, courtesy of Mike Vasile

Game Notes

    On paper, the British should have easily destroyed the American force. However, the lack of British reconnaissance led them into a well-prepared position where they were unable to use their numerical advantages. To reflect the British walking into an ambush, the British need to set up within range of American musketry. With their boats trapped and their cannon unable to successfully deploy, the British have two options. The first is to stay and try to remove the logs blocking the boats from moving further up the creek while taking casualties from the Americans. The other option is to land some or all of their troops and try to frontally assault the prepared defense. If the British player choses this course of action, then any troops that leave the relatively dry boats should spend a turn fording the stream and should not be able to return fire until the following turn. Any cannon fire from the British boats should be ineffective, as the banks of the stream were wooded.

    The fort was a temporary structure. The barrels were placed in either a straight line or V formation and then the captured booty was covered with logs. This not only gave additional protection to hastily constructed defense, but also helped to protect the captured pork and flour barrels. Treat this as light cover.

For game purposes, the British have five units and significantly outnumber the Americans. Moreover, half of the US force are militia, while the majority of the Crown forces are seasoned regulars. If the British can push forward and drive the Americans out before taking too many casualties, they should be able to reclaim the barrels of pork.

The Americans have three units - a small number of elite riflemen, a converged unit of regulars and New York State Militia of dubious quality. The Americans cannot stand up directly to the British, but if they are able to maintain a steady rate of fire and reduce the British forces while retreating to the fort, they may cause enough casualties to keep the British at bay. They can also try to capture the British boats while the British army is off assaulting the "pork barrel fort". The small cannon at the fort itself may play a significant role, but it must be positioned correctly as to inflict the largest number of casualties.

At a scale of 1 to 5, the game plays with 20 American and 50 British figures.

Victory Conditions

Major British Victory: Drive the Americans off the field, capture the "pork barrel fort", destroy the two US gunboats, AND no senior officer becomes a casualty

Minor British Victory: Drive the Americans from the area AND recapture the pork barrels

Minor US Victory: Force the British to leave empty handed

Major US Victory: Destroy the British force AND capture their boats

Further Reading

Beers, Lewis P., *The Pork Barrel Fort, a Documented Account of the Battle of Cranberry/Goose Creek, July 21, 1813*. (Alexandria Bay, NY. Published by the author. 2019)

Hough, Franklin, *A History of Jefferson County in the State of New York*. Joel Munsell. Albany, NY. 1854)

12. British Infantry, courtesy of Richard Watt

Battle of Chateauguay, 26 October 1813

13. Charles de Salaberry, courtesy of John R. Lee

Overview
Battle - 4,000 participants

Outnumbered Canadian militia and Native Americans stop a two-pronged American attack.

Introduction
    For Canadians, the Battle of Chateauguay holds near mythical status, with some calling it as "decisive as Waterloo". While that may be a bit of an over-exaggeration, the battle definitely contributed to the failure of the 1813 American campaign to capture Montreal. Perhaps more importantly, it was the first time an all-Canadian force met the Americans and forced them to retire.
    To stop British supplies reaching Lake Ontario, in the summer of 1813 the Americans planned a two-pronged attack on Montreal. One American Army was to leave from Sackett's Harbor and march northeast and a second smaller force was to advance north from Plattsburgh.
    The campaign was plagued from the start with issues. The US Navy commander on Lake Ontario did not want to risk his ships on an attack against Kingston, so the plan concentrated on the capture of Montreal. The distain between the generals placed in command of two US armies was so great that any communication between the two of them needed to go through the US Secretary of War John Armstrong. As the large American force slowly moved from Sackett's Harbor into the St. Lawrence River, Maj. Gen. Wade Hampton's semi-trained force of 4,000 advanced from Plattsburgh. Capturing the British outpost at Odelltown, Hampton then shifted his forces west to the Chateauguay River. Advancing along the river, the advance guard reported a series of fieldworks a few miles south of the town of Spears.
    The British commander, Lt. Col. Charles-Michel d'Irumberry de Salaberry, was well informed of the American progress and their movement toward Montreal. De Salaberry deployed his troops in three successive lines of defense. The first line was where a creek (English River) flowed into the Chateauguay River. At this location, he created earthworks and placed abatis, rows of cut down trees with their sharpened branches pointing toward the enemy. To guard a ford across the Chateauguay River, about a mile down the river, he posted a second group of men. Finally, another half mile downstream, de Salaberry posted his reserves.

The Battle
    Rather than assaulting the field fortifications directly, Hampton decided to send part of his force under Col. Purdy to cross the river and threaten the Crown forces from the rear. Purdy moved his forces through a swampy waterlogged forest, but his actions were noted and rather than surprising the enemy, he was met by the Canadians previously deployed at this location. As Purdy was stumbling through, Hampton received a poorly phrased order from Armstrong that told him to construct winter quarters for 10,000 men. Believing the campaign to be over, Hampton was ready to withdraw but was interrupted when he heard firing from Purdy's men. As a response, Hampton then sent Izard's brigade forward toward the Canadian defenses. Rather than use his artillery to punch a

hole in the British line, the American Infantry marched into a ravine, deployed into line, and began firing. The fire was ineffective and they did not press the attack. Bugle calls from the Canadians suggested that they were about to launch an assault on the American line. At that point, Hampton became concerned that he was outnumbered and that additional Canadian troops maybe working their way through the forest to envelop the US flank. Rather than continuing to attack in what he believed to be an aborted campaign, Hampton ordered a retreat. Purdy, left to his own devices, managed to extricate himself and rejoin Hampton the next day. The following day the Americans began their withdrawal toward New York State.

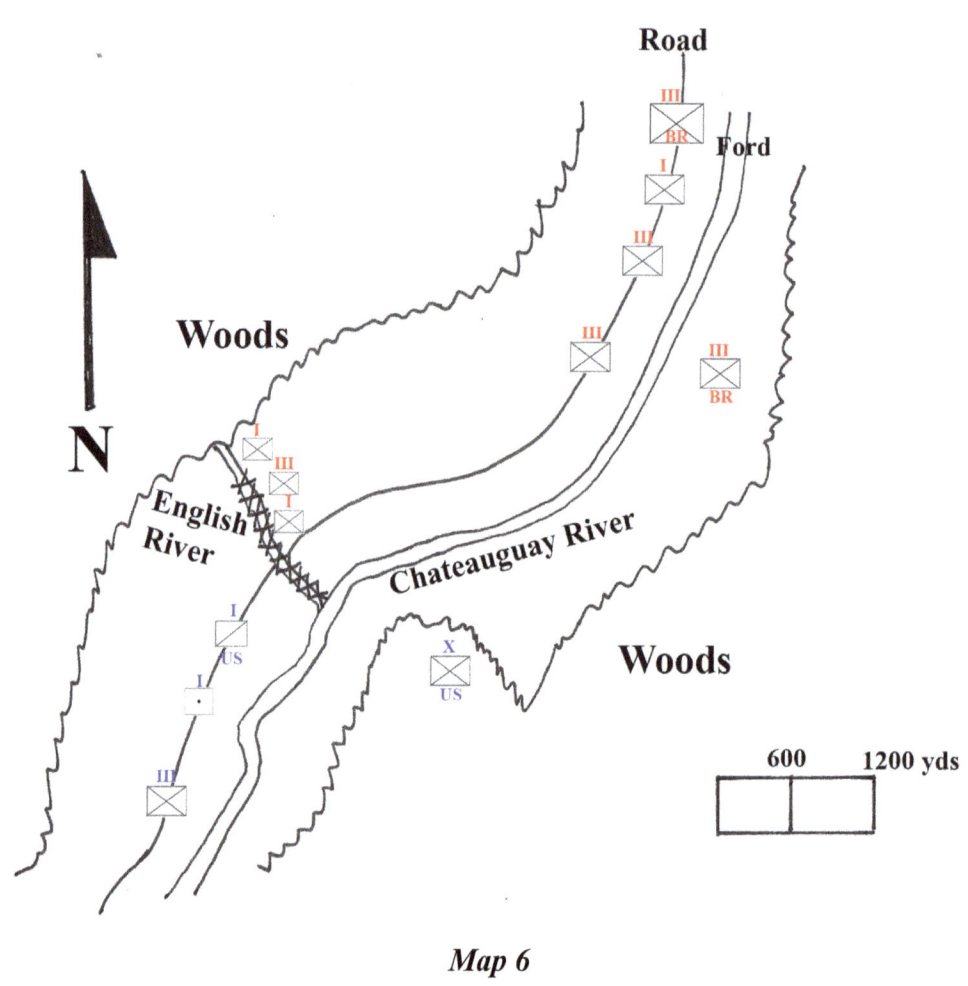

Map 6

Order of Battle

British - 2,000 men
Commanding Officer: Lt. Col. Charles-Michel d'Irumberry de Salaberry

At English River, 408 men
Light Company, Canadian Fencibles, 72 men
Canadian Voltigeurs, 2 companies, 110 men
Select Embodied Militia, 2 companies, 129 men
Sedentary Militia, 75 men
Abenaki and Nippissing warriors, 22 men

At the ford, 160 men
Light Company, 2nd Batt., Select Embodied Militia, *60 men*
Light Company, 3rd Batt., Select Embodied Militia, *60 men*

- 22 -

Sedentary Beauharnois Militia, 1 company, *40 men*
Reserve position, 1,400 men
Canadian Voltigeurs, 2 companies, 110 men
2nd Batt., Select Embodied Militia, 556 men
5th Batt., Select Embodied Militia, 156 men
Sedentary Militia, 4 companies, 220 men
Beauharnois Militia, 4 company, *220 men*
Native warriors, 150 men

US - 3,800 men
Commanding Officer: Maj. Gen. Wade Hamilton

First Brigade 1,697 men
Col. Robert Purdy
Light Infantry Corp, 350 men
Fourth US Infantry Regiment, 591 men
Thirty-Third US Infantry Regiment, 321 men
New Hampshire and Maine volunteers, 415 men
Akwesasne warriors, 20 men

2nd Brigade, 1,737 men
Brig. Gen. George Izard
Tenth US Infantry Regiment, 250 men
Eleventh/Twenty-Ninth US Infantry Regiments, 762 men
Thirtieth/Thirty-First US Infantry Regiments, 700 men
New York State Militia Volunteers, 25 men

Reserve, 350 men
Second US Light Dragoons, 2 companies, 150 men
Two companies of artillery, 200 men
Eight 6-pdr. cannon
One 12-pdr. cannon
One 5.5-inch howitzer

14. Canadian Fencibles, courtesy of Perry Miniatures

Game Notes

The British commander, a French Canadian with previous service in the 60th Royal Americans, a unit of riflemen and light infantry, was the perfect man for the task at hand. Having commanded light troops in Europe, de Salaberry raised and trained a French-Canadian unit of light troops at the start of the war and by 1813, they were battle hardened veterans. Reinforced with a small group of Mohawks, de Salaberry could out scout and out maneuver the Americans. Even the Select Embodied Militia, men who were required to serve for one year, also embraced the light infantry doctrine and emulated their tactics.

The American commander, on the other hand, having received notice that he was to build winter quarters, no longer believed in the endeavor and wanted to resign as the battle was starting. The American troops under Purdy, after marching all night through a swamp, performed as well as possible when they came up onto the prepared Canadians.

Casualties were extremely light for the battle. The British reported two men killed, sixteen wounded while the Americans lost twenty-three men killed and thirty-three wounded. Though one could game the entire battle and have the US forces try to break the three-tiered defense line, another option would be to game the attacks of Izard and Purdy as two separate engagements. Purdy's attempt to flank the British, while brilliant on paper, failed to take into account the terrain. When his brigade had arrived at the predetermined location, all elements of surprise had long since disappeared and his force was exhausted and worn out. Though Purdy tried to advance further, his ability to extract his force makes for an interesting scenario. Movement through the swampy forest was slow and it was impossible to keep unit cohesion. American units should start at one edge of the board and try to successfully withdrawal as many companies as possible within a given number of turns. The US player has the ability to stand and trade volleys with the Canadians, or even attempt to melee, but then loses the ability to move for that turn.

Gaming Gen. Izard's actions is difficult to replicate on the table. Though the plodding attack, made without artillery support, can be duplicated, the fear of a British flanking attack from the wood line is problematic to replicate. It is possible to modify the original disposition of forces and have additional British troops appear on the American flank. If such a plan of attack is followed, I would allow the US to bring up a cannon or two to make holes in the Canadian defenses. The game then becomes a more "what-if" scenario, but it makes for a much better tabletop activity. Players are allowed to deploy as they see fit within their respective deployment zones as indicated on the map.

15. US Infantry, author's collection

Victory Conditions

The Americans need to drive the forces of the Crown off the field to achieve a total victory, but feel free to use your rules to set determine victory conditions based upon casualties and unit loss.

Further Reading

Elting, John R., *Amateurs, To Arms!: A Military History of the War of 1812.* (Da Capo Press. New York. 1995)
Graves, Donald E., *Field of Glory: The Battle of Crysler's Farm, 1813.* (Robin Brass Studio, Inc. Montreal. 1999)
Izard, George, The War of 1812 in Northern New York: General George Izard's Journal of the Chateauguay Campaign. (*New York History.* April 1995: 173-200. 1995)
Wohler, J. P., *Charles de Salaberry: Soldier of the Empire, Defender of Quebec.* (Dundurn Press. Toronto. 1984)

Crysler's Farm, 11 November 1813

Overview
Battle - 5,000 participants
Small Skirmish - 450 participants

American troops try and fail miserably to dislodge a smaller British force that threatened their rear. The highlight of the battle is a small desperate charge by a company of cavalry.

Introduction
In the early fall of 1813, American forces had launched a much-delayed advance towards Montreal. By capturing the city, the Americans hoped to cut off the British and Canadian troops in Upper Canada from British supply lines. To achieve this goal, the Americans created two separate armies – one led by Maj. Gen. Wade Hampton and the other by Maj. Gen. James Wilkinson. Hampton advanced from the American base at Plattsburgh along the Chateauguay River, while Wilkinson moved up from Sackett's Harbor along the St. Lawrence River. The two forces were to link up at Montreal.

Coordination between the two rival generals was poor and following the American defeat at the Battle of Chateauguay (26 October), Hampton withdrew back to Plattsburgh. Wilkinson continued his advance, still hoping to be reinforced by Hampton. To deal with the American threat, a small force, designated a "Corp of Observation" was dispatched under Lt. Col. Joseph Morrison from Kingston in pursuit of the invaders. After crossing the St. Lawrence River into Canada, the Americans bypassed Ft. Wellington and were approaching the Long Sault – eight continuous miles of extremely dangerous rapids that could only be traversed during daylight. On the 10th, Gen. Brown assumed command of an advanced guard and routed the British garrison at Ft. Matilda, which guarded the river's narrowest point, while Brig. Gen. Boyd commanded the rear guard. That evening, Wilkinson, ill from dysentery, took to bed and Maj. Gen. Lewis assumed temporary command of the US forces.

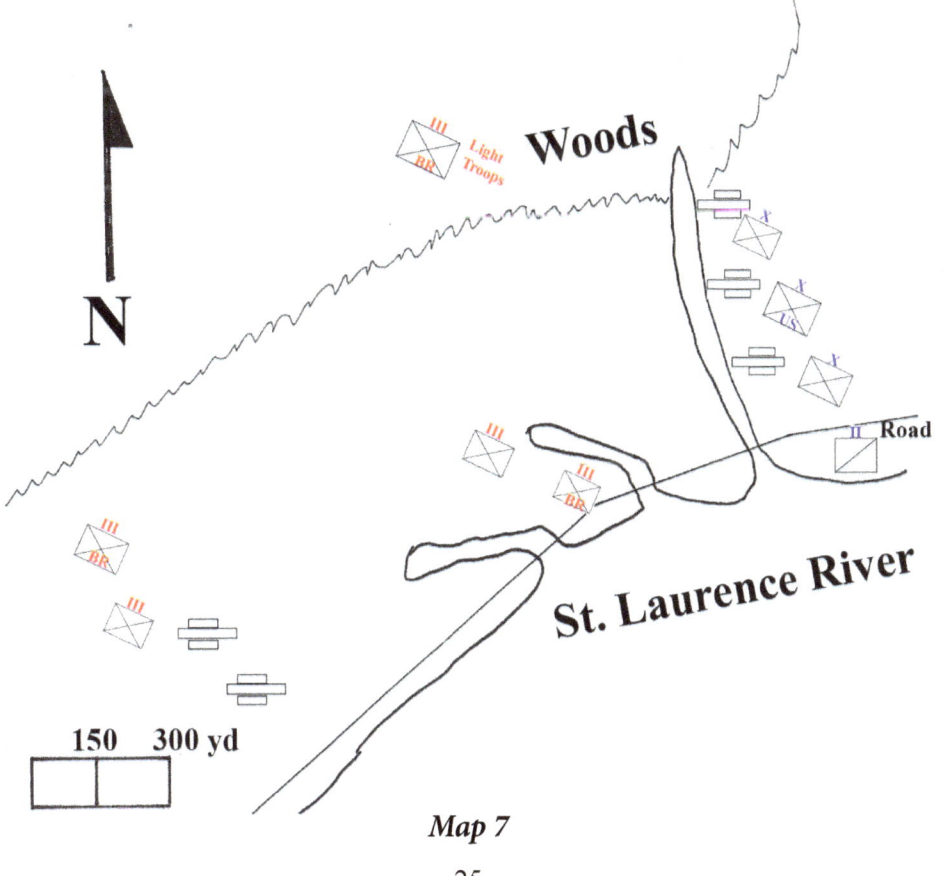

Map 7

On the morning of the 11th, Wilkinson unexpectedly reassumed command and changed all of Lewis's orders. Wilkinson ordered an advance through the rapids, but when Gen. Boyd warned of an imminent enemy attack, Wilkinson then cancelled the advance and ordered Boyd to take his brigade and drive the Canadians off. At this point, Wilkinson returned to his bed and took no further action in the battle.

Noting the US troops were planning to attack, Morrison, now reinforced with men from Ft. Wellington, went on to the defensive and took up a position on the narrow western end of John Crysler's Farm. With the British left flank covered by a swampy forest and his right by the St. Lawrence River, where he could be supported by British gunboats, Morrison placed his 3 guns in the center, which were supported by his regulars.

The Battle

16. John de Barth Walbach (Ft. Monroe, 1846)

Boyd's advance was poorly planned and thus, failed. Troops were committed piecemeal and it was only later in the battle that any artillery was brought forward. The initial American advance drove the British skirmishers back, but when Gen. Covington suffered a mortal wound and his second in command was killed shortly thereafter, his brigade began to retire and Gen. Swartwout's men began to follow. As the American line faltered, Morrison advanced, threatening to capture the American guns. A single squadron of American dragoons, led by Col. John de Barth Walbach, Wilkinson's adjutant, then charged the exposed British flanks. Pummeled by artillery fire from the British gunboats, the three cavalry charges made over muddy broken ground failed to close with the enemy, but forced Morrison to stop and refuse his flank, which gave time for the Americans to withdraw their guns. As the Americans gradually fell back to the river, the battle came to an end.

The following day, the Americans crossed the rapids uneventfully. Wilkinson then received news that Hampton had gone into winter quarters and following a council of war, withdrew. The campaign was over.

Order of Battle

British - 1,189 men
Commanding Officer: Lt. Col. Joseph W. Morrison
49th Regiment of Foot, 8 companies, 304 men
89th Foot, 5 companies, 240 men

The Advance, 186 men
Lt. Col. Thomas Pearson
Grenadier company, 49th Foot, 39 men
Light company, 49th Foot, 39 men
Canadian Fencibles, 2 companies, 108 men

Capt. Barnes's Command
Capt. George West Barnes
89th Foot, 3 companies, 144 men

The Light Troops, 192 men
Maj. Frederick G. Heriot
Canadian Voltigeurs, 3 companies, 150 men
Provincial Dragoons, 12 men
Native warriors, 30 Mohawks

Royal Artillery, 63 men
three 6-pdr. guns

Dundas Militia, 60 men (present at the battle but were not engaged in combat)

British flotilla - Two schooners and seven gunboats. The schooners could not make it further upriver and returned to Kingston.

U.S. - 2,450 men
Brig. Gen. Boyd, commanding
First Brigade – Col. Isaac Coles
    Twelfth US Infantry, 225 men
    Thirteenth US Infantry, 225 men
Third Brigade – Brig. Gen. Leonard Covington
    Ninth US Infantry, 300 men
    Sixteenth US Infantry, 225 men
    Twenty-Fifth US Infantry, 375 men
Fourth Brigade – Brig. Gen. Robert Swartwout
    Eleventh US Infantry, 300 men
    Fourteenth US Infantry, 125 men
    Twenty First US Infantry, 425 men
Col. John de Barth Walbach
    Second Regiment of US Light Dragoon, 1 sqn, 150 men
Artillery – six 6-pdr. guns, 100 men

Gaming the Battle

    I have gamed this overall battle at different times with different outcomes. The 2 ½ to 1 odds in favor of the Americans suggests that the engagement should have gone to the US, but the Americans were hampered by poor leadership and the loss of a brigade commander and then his second in command. Also, being fired upon by the British gunboats should not be discounted.

    As to the generals, Boyd clearly was promoted beyond his abilities and was not up to the task assigned to him. Morrison, however, demonstrated competency, though risking his small force bordered on reckless and he was lucky he faced an incompetent adversary. Had he faced a competent general, Morrison may have encountered overwhelming odds and had his force destroyed. Though the British suffered far fewer casualties than the Americans, Morrison was in no position to exploit his victory and pursue the enemy.

    As to the junior officers, Col. John de Barth Walbach, Wilkinson's adjutant, stands out. A Catholic aristocrat from Munster, this native German speaking Hussar officer, after serving for twelve years in French, Prussian and Austrian armies against Revolutionary France, was promoted to Major in the emigre Hussars de Rohan. Following his service with the British army in Santa Domingo, he followed his elderly father to the US. Walbach's actions at Crysler's Farm are an embodiment of the ideals of the 19th century – bravery bordering on foolishness for the sake of glory. Not only were his three charges made over broken ground in the face of artillery fire, but for the purpose of saving the guns. Half a century later, Lord Cardigan would be glorified by the poet Alfred Tennyson for leading a somewhat similar action in "The Charge of the Light Brigade" in the Crimean War.

17. Canadian Voltigeurs, courtesy of Sash and Saber Casting

Gaming the Cavalry Charge

For those who want to game just a section of the battle, recreating Col. Walbach's charge can make for an interesting game. When the Americans finally brought up their guns, the US Infantry was already retreating. Placed along the river road, the guns began to pound the British. Lt. Col. John Harvey, Morrison's second in command, was ordered to capture the guns. Crossing the muddy ground and crossing a number of split-rail fences, the British moved forward. As they emerged, the 49th exposed their flank. It was at this moment that the dragoons moved forward. The British troops were able to wheel and fire on the dragoons. Though the cavalry never came into contact with the enemy, their actions allowed the US to withdraw five of the cannons, while the sixth, which had got immobilized in the mud, was abandoned.

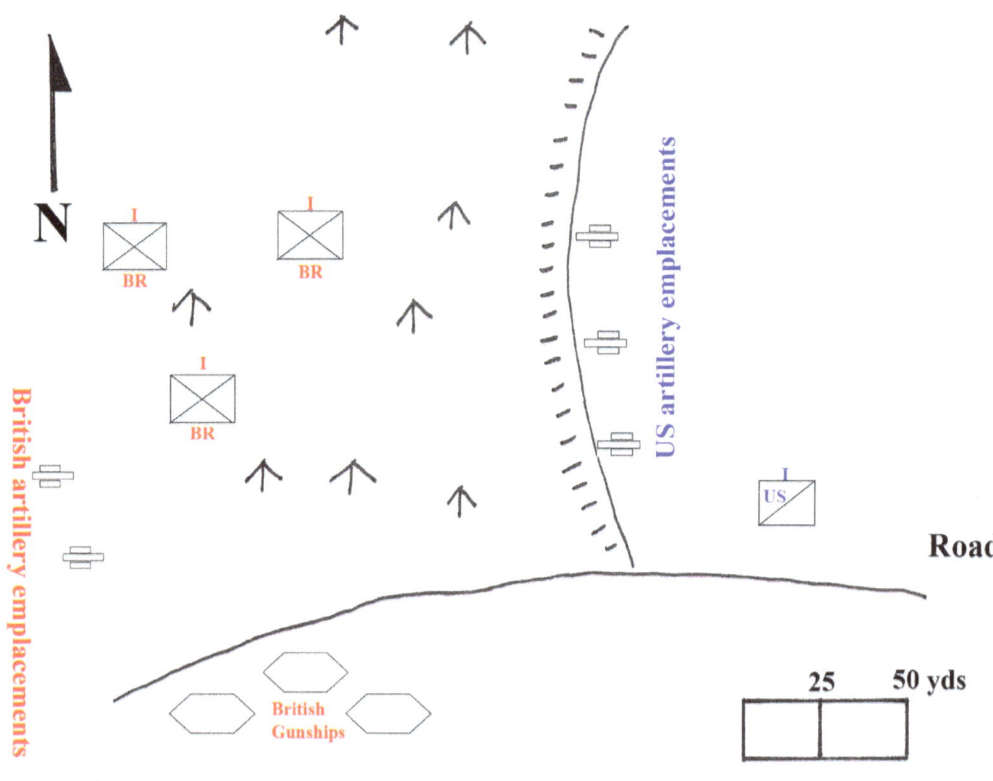

For this action, the forces should include:

US
Col. Wabach
1 company of dragoons, 130 men
6 guns and gunners, *50 men*
Handful of US troops skirmishing near the guns, *20 men*

British
Lt. Col. John Harvey
49th Foot, 2 companies, *120 men*
Detachment of Canadian Fencibles, 150 men
Two 6-pdr. guns
Artillery fire from the supporting gunboats

Map 8

Scenario Gaming Notes

When trying to move the guns, the rules should require a die role to account for the poor conditions. Since the mud effected both sides equally, movement for all should be reduced. For the British formed troops crossing the fences, each time they reach a fence, they should halt. After crossing the fence, next movement phase should require them to form up before moving forward. Given the small number of troops involved with this actions, one can do it at a scale where 5 men are represented by 1 miniature.

Troop Ratings

For the most part, the American Infantry behaved poorly during the battle. In part it was the cold weather that effected the performance of the underfed southern troops, many of which were still in their summer uniforms, but a far more critical component was the poor leadership. Generals Wilkinson, Lewis, and Boyd were clearly not up to the task, while the death of the Third Brigade's two commanders effectively neutralized that unit. Harvey, on the other hand, demonstrated both his bravery and tactical knowledge by leading the despite cavalry charges. Had General Brown or Porter been put in command, the American attack would have gotten off to a much better start, but Boyd was, at best, unsuited.

The British troops were led by professionals and behaved accordingly. Tactically, Morrison emulated Wellington almost exactly. Morrison first picked a piece of ground suitable for defense and let the enemy come to him. Once the enemy attack had been broken up by artillery fire from both gunboats and the field artillery, the British went on a counter-offensive. The elite companies of the 49th were able to wheel under fire and face the oncoming cavalry charge. The Canadian Fencible Regiment had been in service for a number of years and was no longer a "militia", but a well-trained British unit. The Mohawks and Canadian Voltigeurs did what was expected of them.

18. US Light Dragoons, courtesy of Knuckleduster Miniatures

Victory Condition

For a side to claim a brilliant victory, they must drive the enemy from the field while their entire command suffers no more than 30% casualties.

For those wanting to game just the cavalry action, the American player's goals are to save the guns. If the American player is able to remove all six guns to the rear, they have a stunning achievement. If they are able to save five, four or three guns, it's a partial US win. If the British are able to stop the Americans from achieving this goal, they win the engagement.

Further Reading

Elting, John R., *Amateurs, To Arms!: A Military History of the War of 1812.* (Da Capo Press. New York. 1995)

Graves, Donald, *Field of Glory: The Battle of Crysler's Farm*, 1813. (Robin Brass Studio, Inc. Montreal. 1999)

Way, R. L., *The Day of Chrysler's Farm.* (St. Lawrence Parks Commission. Morrisburg, Ontario. 1960)

19. 49th Foot, courtesy of Roger Chrysler

Lacolle Mill, 30 March 1814

Overview
Battle – 5,000 participants

In order to redeem his reputation, US Maj. Gen. Wilkinson launches an advance to save his career. His failed assault is the last US attempt into Lower Canada.

Introduction
Following the disastrous 1813 US campaign against Montreal, Maj. Gen. James Wilkinson boasted in November 1813 to the Marquis de Lafayette that "Our quarters for this winter will probably be Montreal". Though Wilkinson dreamed of launching offensives against Kingston, Prescott, or even Quebec, he did not have the troops, supplies, or transportation to do so. Wilkinson sought action, as he was aware that he was likely to be relieved of command due to his mismanagement of the attack on Montreal. Learning that an isolated British garrison just north of the American border had converted a stone mill into a British outpost, Wilkinson launched an attack to capture Lacolle.

Campaigning in winter, especially in Canada, is difficult at best but Wilkinson pressed forward. Before marching out, Wilkinson boasted to "return victorious or not at all". With snow still deeply laying on the ground, progress was slow. Brig. Gen. Alexander Macomb's brigade was sent forward to prevent any relief force from the garrison at Ile aux Noix from reaching the so-called "isolated" outpost.

The British outpost at Lacolle consisted of an old stone mill converted for military use, recently built barracks, and a blockhouse. With its 18-inch-thick-walls, the mill was a tall and solid structure on the south side of the river. The mill's windows were barricaded with strong timber and were cut with loopholes for muskets. On the other side of La Colle River stood a two-story wooden blockhouse and around them were log breastworks. Manned by British Regulars, Royal Marines and Canadian Voltigeurs, the mill was supported by British troops garrisoning *Ile aux Noix*, a few miles away.

The Battle
After making a number of wrong turns and floundering in the deep snow for thirty miles, Wilkinson reached the mill. He then sent his advance guard, along with Smith's regiment, to flank the outpost from the rear and cut off the enemy's retreat.

The main American force then moved to within 150 yards of the building and opened fire, first with muskets and then with artillery. The ground was too soft to support their 18-pdr. cannon, so Wilkinson tried to use his two 12-pdr. cannon and his 5.5-inch mortar. Neither had any noticeable effect on the outside of the structure but given the amount of kinetic energy released when a twelve pound

20. La Colle Mill and Blockhouse, (Lossing)

iron solid shot cannon ball impacts a stone structure, the damage the British troops suffered when taking shelter in the building must have significant[2]. The British replied with Congreve rockets, which wounded both Captain McPherson, the commander of the guns, and his second in command, Lt. Larrabee.

With the arrival of the two flank companies of the 13th Regiment of Foot from nearby Isle aux Noix, the British attempted to storm the American guns. Repulsed, they made several additional charges, which were also rebuffed by the US Infantry of Smith's and Bissell's Brigades. Eventually driven back to the blockhouse, the British were reinforced by additional Canadian grenadiers and Voltigeurs from nearby Burtonville. The British then launched another desperate sortie against the Americans and were again repelled.

By this time, the British garrison at the mill was almost out of ammunition. The Royal Navy, in the form of a sloop and several gunboats, arrived on the scene and began to bombard the American line from the Richelieu River. Unable to breach the mill by artillery fire, Wilkinson hesitated. The British forces, continually reinforced, now numbered upwards of a thousand men. The Americans had suffered over 150 casualties, including its artillery commanders. Concerned about the weather and coming darkness, Wilkinson ordered a retreat back to Plattsburgh. On 11 April Wilkinson was relieved of command.

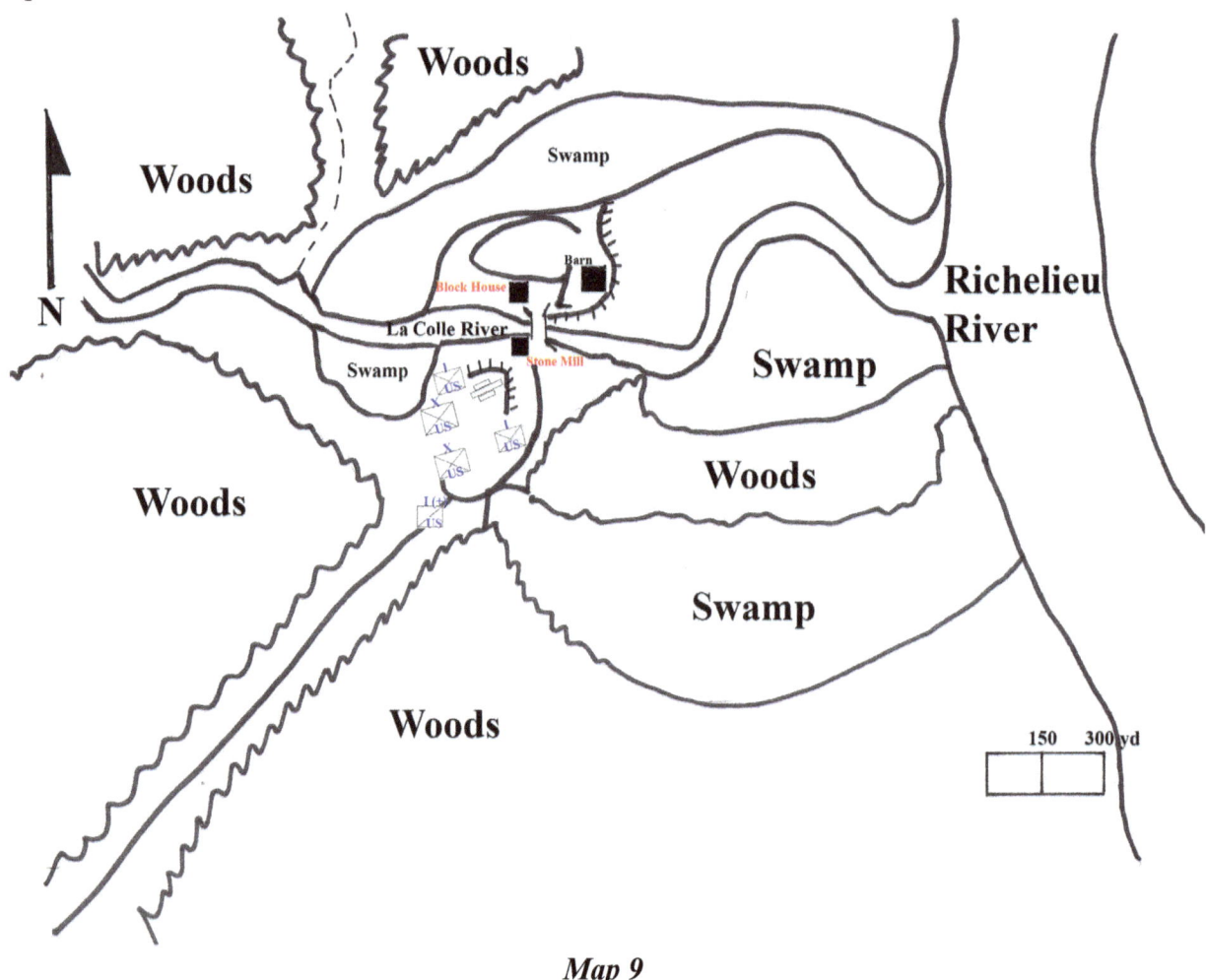

Map 9

Order of Battle

British - 240 at the outset of the battle, upwards of a 1,000 at the end
At the Mill
Commanding Officer: Maj. R. B. Handcock

---

[2]The ball is about 5 kg, with a velocity of about 350 meters/second. A half-meter stone wall absorbing the iron ball over a time of 0.1 seconds calculates out to 613.5 kN (kilo Newtons) with a peak impact of 1,225 kN. That is equivalent to a 135 ton blow. This would shatter stone and masonry, sending rock spalls and wood splinters everywhere. It is not surprising that the British would rather charge the American line than continue to be pounded by the American artillery.

13th Regiment of Foot, 60 men under Capt. Blake
Royal Marines, 70 men
A Detachment of the Royal Marine Rocket Corp, 30 men
Canadian Voltigeurs, 80 men

From Burtonville
Capt. Cartwright
Canadian Fencibles, 80 men
Frontier Light Infantry (English volunteers in the Canadian Voltigeurs), 80 men under Capt. Ritter
Mohawks, 40 men

From Ile aux Noix, arriving shortly after the battle began
Grenadier company, 13th Foot, 80 men under Capt. Ellard
Light company, 13th Foot, 80 men under Capt. Holgate

From Ile aux Noix, arriving near the end of the battle
A Sloop and several Gunboats, Capt. Pring

US - 4,000 men
Commanding Officer: Maj. Gen. James Wilkinson
The precise order of battle for the US forces cannot be established. For gaming purposes, however, the following hypothetical organization was pulled together from incomplete, and often conflicting, sources.

Advance Guard, 600 men
These troops were sent to cut off the enemy's retreat and did not participate in the assault on the mill.

Col. Isaac Clark and Maj. Benjamin Forsyth
Eleventh US Infantry, *1 company, 60 men*
30th Infantry Regiment, *250 men*
31 Infantry Regiment, *250 men*
First US Rifle Regiment, *detachment, 40 men*

First Brigade, *1,100 men*
Brig. Thomas Adams Smith
Forth/Tenth Infantry Regiment, Col. Purdy, 600 men
Twelfth Infantry Regiment, 500 men
Two 6-pdrs

Second Brigade, *1,000 men*
Brig. Daniel Bissell
Fourteenth Infantry Regiment, *350 men*
Twentieth Infantry Regiment, *350 men*
Twenty-third Infantry Regiment, *300 men*
Two 6-pdrs

Reserve, *1,000 men*
Brig. Alexander Macomb, did not participate in the action mandy was held in reserve
Fifteenth US Infantry, 300 men
Sixteenth US Infantry, 300 men

Twenty-first US Infantry, 300 men
US Light Dragoons, 100 men

Artillery, *304 men*
Capt. Robert Macpherson
Artillerists, 304 men
11 guns, including one 18-pdr. gun, one 12-pdr. gun, and one 5.5-inch mortar.

21. British Rocket Battery, courtesy of Richard Watt

Game Notes

Having gamed the entire battle a number of times, both sides have a number of liabilities. For the British, being outnumbered, surrounded, and with limited ammunition, their choices are rather limited. If the Americans are able to damage the stone mill, which would happen given enough time, or summon up the courage to assault the mill directly, the center of the British defense line would crumble, and the Crown forces would be pushed back to the blockhouse. Then, with the path cleared to bring up the guns, the Americans could destroy the blockhouse and with nowhere to run, the British would be forced to surrender.

The British can either remain on the defensive or try and sortie out to destroy the American guns. If they are successful in capturing and holding the guns, the British would earn a glorious victory. The other option is to capture and spike the guns, which if successful, would destroy the American ability to batter down their defenses. Finally, the British can still win by remaining on the field.

Though the Americans outnumber the British three to one (Macomb's Brigade did not take part in the action and the advance guard may have never secured the rear), time is against them. Wilkinson is a poor commander and never pressed his attacks. Furthermore, by dispersing his forces, it is difficult for the Americans to achieve local tactical superiority. If using the historical placement of US forces, it is difficult to make full use of the American numbers.

Given the March weather in Lower Canada, troop movement should be slow. As the melting snow begins to thaw, in places the ground becomes mush. For this reason, before the American player can place a gun, they need to roll a die. The heavier the gun, the less likely that it can be placed in a certain location. Thus, a die roll of a 6 is required to place the 18-pdr. gun, a 5 or a 6 for a 12-pdr. gun, 4, 5, or 6 for the howitzer and 3, 4, 5, or 6 for the lighter pieces.

Congreve rockets, though historically inaccurate, were known to cause havoc and often would result in an enemy fleeing the field. At this engagement, the American line held (it passed its morale check) even after both the US artillery officers assigned to this advance were hit by the "infernal gadgets".

For the Americans to win, they need to press their attacks quickly, especially as the British were receiving reinforcements. To add a bit of chance to the battle, one should roll for reinforcements, including when and where they come on the table. One should also not limit themselves to the actual British troops that took part. In one game played recently, the American line collapsed when a small company of Canadian militia showed up on its southern flank.

Finally, for the Americans, high casualties are to be avoided. Wilkinson wanted a quick victory and if too many lives were wasted on capturing the isolated outpost, his political enemies would highlight his wasting of men on an unimportant objective. Thus, once the American army suffers more than 10% casualties, they should immediately begin to withdraw.

Victory Conditions

The Americans must capture and hold the mill AND the blockhouse at the end of turn eight. The British win if they prevent the Americans from achieving their victory conditions. Consult your favorite rules to use limited ammunition.

Further Reading

Elting, John R., *Amateurs, To Arms!: A Military History of the War of 1812.* (Da Capo Press. New York. 1995)

Gosling, D. C. L., The Battle of LaColle Mill, 1814. (*Journal of the Society for Army Historical Research*. Vol. 47, No. 191, Autumn. 169-174. 1969)

Lossing, Benson J., *Pictorial Field-Book of the War of 1812.* (Harper & Brothers Publishing. New York. 1869)

Wilkinson, James, *Memoirs of My Own Times*. (Printed by Abraham Small. Philadelphia. 1816)

22. US Artillery, courtesy of Knuckleduster Miniatures

Bad Roads and Poor Rations

Plattsburgh, September 1814

Overview
Naval Engagement
Large Battle - 15,000 participants
Battle - 4,600 participants
Small Skirmish - 360 participants

The British Army invades and captures Plattsburgh, but an American naval victory does not allow them to establish themselves in the area and they withdraw.

Introduction

23. Thomas MacDonough by Thomas Gimbreder. (Naval History & Heritage Command)

With the abdication of Napoleon in April of 1814, The British government had at its disposal a large army with no one to fight. With the input of the Duke of Wellington, the British sent the veterans of the wars against Napoleon to North America. So many British troops were sent to the Montreal and Quebec area that they were outstripping the available food stores. Sir George Prevost, the British Governor and Commander in North America, wanted to launch a decisive attack on Sackett's Harbor and in this one swift action, cause a collapse of the American presence on the Great Lakes. However, to stock the necessary supplies at Kingston, the logistical build up required the roads between Montreal and Kingston to freeze. Thus, the offensive would only start in the spring of 1815, something that neither the British nor Canadian governments wanted to wait for. As a compromise, Prevost began planning a joint Army/Navy operation on Lake Champlain.

The Lake Champlain corridor was a major theater of military operations since the French and Indian War and served as a conduit for the British during the disastrous Saratoga campaign of 1777. During the first years of the War of 1812, British had raided down the lake with impunity and American cattle from the area found its way into the stomachs of British troops.

Aware of the regional British troop build-up, US Secretary of War John Armstrong ordered the recently promoted regional commander Maj. Gen. George Izard to advance toward the St. Lawrence. After replying that he would carry out the Secretary's orders, but he had no hope of success, Armstrong ordered Izard to march across New York State to Sackett's Harbor, taking 4,500 troops with him.

Before departing west, Izard, a military engineer, constructed a series of fortifications south of Plattsburg. The Americans also went on to strengthen their naval presence on the lake, building the twenty-six gun sloop of war *Saratoga*, the twenty gun brig *Eagle*, converted the hull of a steamboat into the seventeen gun *Ticonderoga*, and rehabilitated ten gunboats with Columbiads, guns that fired fused explosive ammunition.

Prevost began his campaign on 31 August and was making good progress. The British inland column, when ambushed by a thousand American troops sent out to harass them, did not deploy into line, but advanced in a march column against the enemy. The Americans retired, only to set up a second ambush down the road. Here again, the Americans made another successful stand, but then withdrew when pressed by superior numbers. The

American gunboats, deployed in Plattsburgh Bay, then began to shell the British column as it advanced. With the coming British force, the Americans withdrew south across the Saranac River, taking up the flooring of the two bridges as they went across, and deployed in the previously prepared fortifications designed by Izard.

The Battle

Having secured the town of Plattsburgh, Prevost noted the presence of a significant American force occupying the high ground just south of town. Entrenched and supported by blockhouses, the American line was reinforced by the American squadron in Plattsburgh Bay. Rather than risk an all-out assault against a prepared position, Prevost decided to wait for the arrival of the Royal Navy. As Prevost waited, more American militia arrived to reinforce the fieldworks. On the night of 9 September, fifty Americans under Capt. George McGlassin slipped over the Saranac River and destroyed a British Congreve rocket battery.

Prevost's plan of attack was simple. Once the Royal Navy had made short work of the American flotilla, one British brigade would make a feint against the two bridges, while two additional British brigades would ford the Saranac River a mile upstream and then attack the American line from the west.

On the morning of the 11th, the Royal Navy under Captain George Downie sailed into the bay and Prevost launched his assault. The British were unprepared for the assault, as some regiments were still eating breakfast. By 10:00 AM, the two brigades entrusted with outflanking the Americans made a wrong turn, which took an additional hour to correct. The American guns dueled with their British counterparts and those British troops entrusted with making a feint against the American line made no visible progress.

In the Bay, the American squadron devastated the British assault. Downie was killed at the start of the engagement. Maximizing the short-ranged guns of the American squadron and loosely laid anchor cables, which gave the American vessels the ability to turn quickly and bring into action their previously unused broadsides, the Royal Navy was soundly defeated. With most of the British warships surrendered, the British gunboats retreated. When news of the American naval victory raced through the lines, Prevost hesitated in pressing forward. Without naval support, the British occupation of Plattsburgh could not be sustained or supplied over the coming winter. Since destroying the American troops would not serve any larger strategic goals, the potentially bloody assault was suspended. Over the course of the afternoon, the artillery fire from both sides began to subside until it eventually fizzled out. Once darkness fell, Prevost began to withdraw his troops. A small token American force of light troops followed the British, but these soon returned, fearful of a possible counterattack by the much larger British Army. Returning to Canada, the failed assault cost Prevost his career.

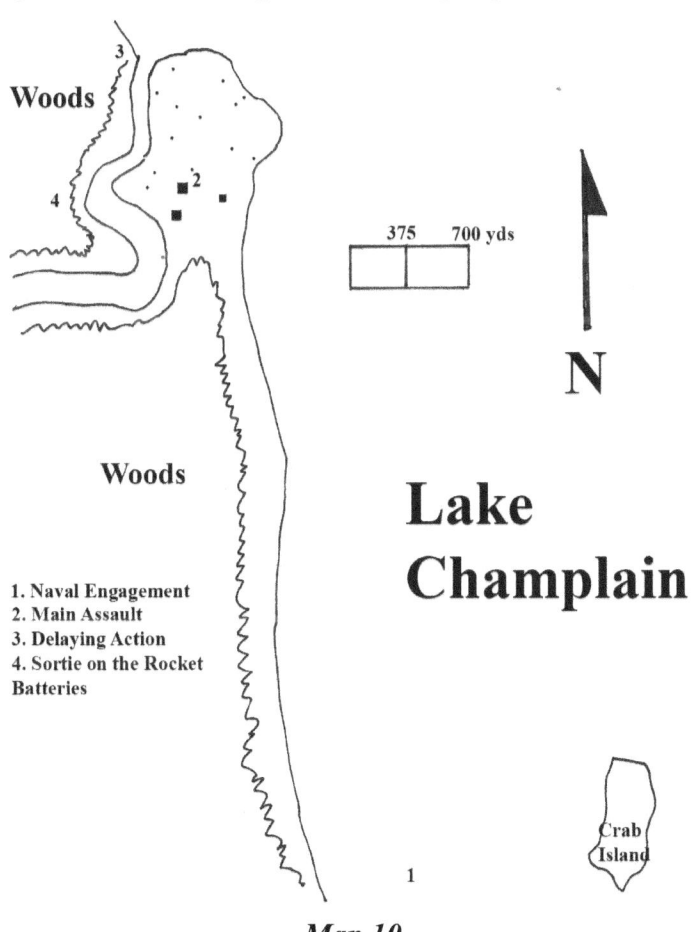

1. Naval Engagement
2. Main Assault
3. Delaying Action
4. Sortie on the Rocket Batteries

Map 10

Game Notes

Prevost is often portrayed as an incompetent military commander, but he was well aware of the risks in attacking prepared American defenses following his involvement in the 1813 British attack on Sackett's Harbor.

It has been also argued that Prevost should have pressed his attack on the Americans immediately following his capture of Plattsburgh, but apparently, he listened to the advice of Robinson, whose brigade was to lead the assault and had taken casualties on their way to Plattsburgh. Prevost chose to wait for the Royal Navy to silence the guns of the American flotilla.

Scenario #1 - Naval Engagement

For those enjoying naval campaigns, the battle between the two squadrons makes for an interesting engagement. The British flagship, The HMS *Confiance*, though practically bigger than the entire combined US flotilla, was recently built and poorly crewed. Though the British boats had longer ranging guns, Downie ordered his squadron to close on the Americans, thus eliminating his advantage in range.

In duplicating the naval engagement, both Crab Island and the coast are fringed with shallows. The American commander knows where they are, while the newly dispatched British Captain does not. If a British warship (not gunboat) comes within 12 inches of either the island or the mainland, there is a 1 out of 3 chance they will run aground. If a British warship does run aground, there is a 1 out of 3 chance they can free themselves the following turn. Ships choosing to free themselves may not take any additional actions that turn.

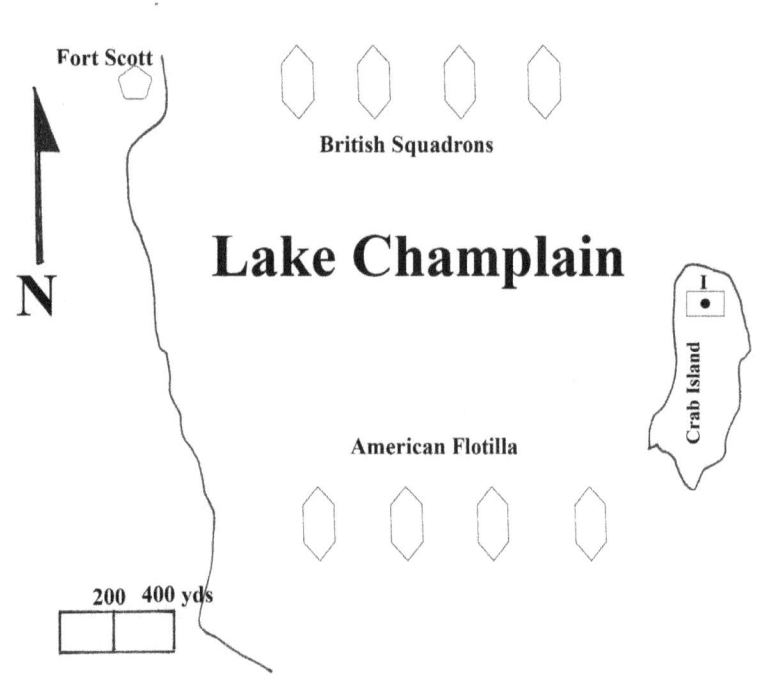

Map 11

Royal Navy
Capt. Downie 1,000 Men
HMS *Confiance*, Fifth Rate Frigate, 37 guns - thirty-one 24-pdrs., six 32-pdrs.
HMS *Linnet*, Brig, 16 guns - sixteen 18-pdrs.
HMS *Chubb*, Sloop, 11 guns - one 6-pdr., ten 18-pdr.
HMS *Finch*, Sloop, 11 guns - four 6-pdrs., seven 18-pdr.
Twelve Gunboats, eight with two guns, four with one gun

US
Thomas Macdonough – naval forces, 882 men
USS Saratoga, Corvette, 26 guns - eight 24-pdrs., six 42-pdr. carronades, twelve 16-pdr. carronades
USS Eagle, Brig, 20 guns - eight 18-prds., twelve 32-pdrs.
USS Ticonderoga, Schooner, 17 guns - four 18-pdrs., eight 12-pdrs., five 32-pdr. carronades
USS Preble, Sloop, 7 guns – seven 9-pdrs.
10 Gunboat Galleys
     six mounting one long 24-pdr. and one 18-pdr. Columbiad
     four with one 12-pdr.

On Crab Island - a two 6-pdr. gun battery manned by invalids from various regiments

Victory Conditions

The Royal Navy must sink the American fleet to attain victory. If the American squadron remains on Lake

Champlain, they remain a threat to the massive British army.

Scenario #2 – Main Assault

It is also possible to game the British assault on the American works, either immediately following the British capture of Plattsburgh, or as when the Royal Navy engaged the American flotilla. The large size of the British land force, along with the formidable American defensive lines, makes for a balanced game. If gaming the "what if" variant - Prevost launches a direct assault before the arrival of the British squadron - The British must directly assault the US in front of them and not diverge their forces to find and turn a flank. This reflects the low opinion the British had of the American "farmers". At the same time, the American player should reduce the amount of militia available by half but be able to count on naval artillery support. If your rule set does not have specific rules for naval supporting fire, treat each ship as a gun/battery firing at long range.

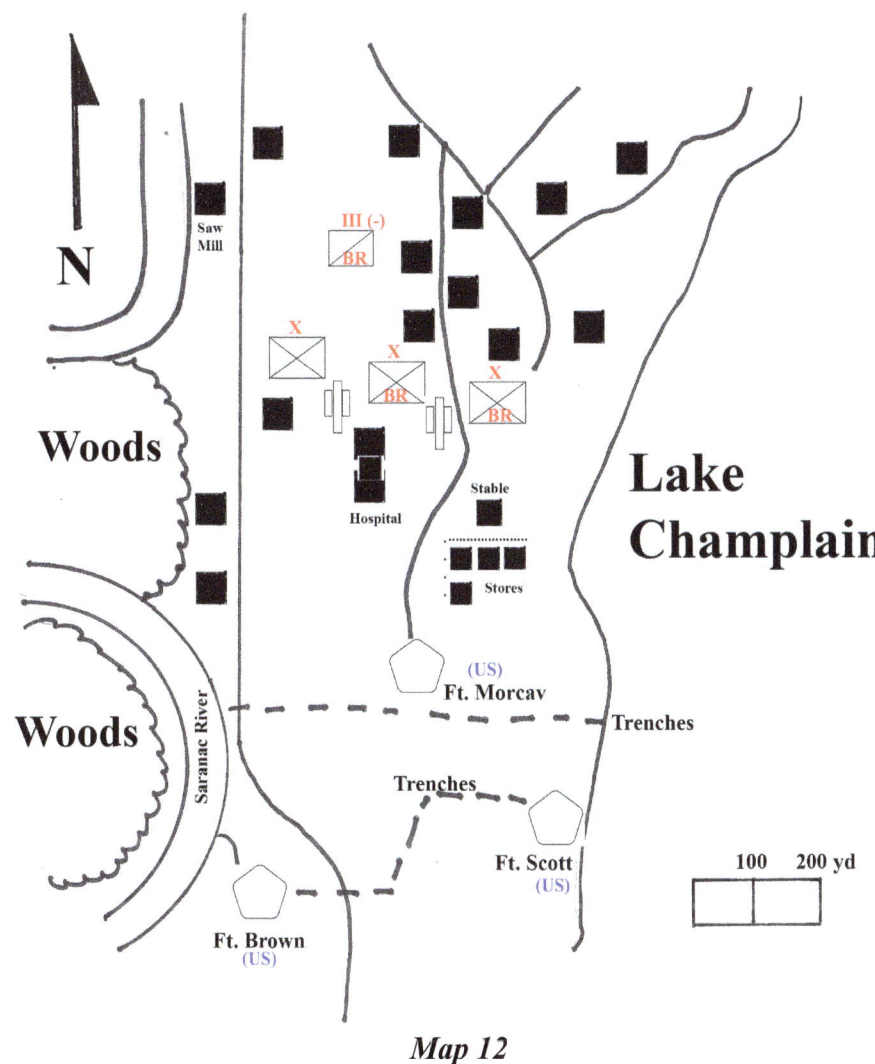

Map 12

Order of Battle

British – approximately 9,000 men (not including the Royal Navy)
Commanding Officer: Lt. Gen. Sir George Prevost

1st Brigade - *1,500* men
Maj. Gen. Frederick Robinson
3rd Batt., 27th Regiment of Foot
1st Batt., 39th Regiment of Foot
76th Regiment of Foot
1st Batt., 88th Regiment of Foot
Royal Artillery field battery - five 6-pdr. cannon and a 5.5-inch howitzer

2nd Brigade – *3,200* men
Maj. Gen. Brisbane
2nd Batt., 8th Regiment of Foot
13th Regiment of Foot
49th Regiment of Foot
De Meuron's Regiment
Canadian Voltigeurs
Canadian Chasseurs
Royal Artillery field battery - five 6-pdr. cannon and a 5.5-inch howitzer

3rd Brigade – *3,500* men

Maj. Gen. Manley Power
3rd Regiment of Foot
5th Regiment of Foot
1st Batt., 27th Regiment of Foot
58th Regiment of Foot
Royal Artillery field battery - two 6-pdr. cannon and a 5.5-inch howitzer

In addition, two squadrons of the 19th Light Dragoons and a small siege train of artillery was attached to the Division. This included two 24-pdr. brass field guns, an 8-inch brass howitzer, three 24-pdr. naval carronades mounted on field carriages and a detachment of Congreve rockets.

US - approximately 5,000 men (not including Naval forces)
Commanding Officer: Brig. Alexander Macomb

Thirteenth U.S. Infantry Regiment, 200 men
Fifteenth U.S. Infantry Regiment, 100 men
1,500 from the Sixth, Twenty-Ninth, Thirtieth, Thirty-First, Thirty-Third, and Thirty-Fourth U.S. Infantry Regiments
50 Artillerymen
700 New York State Militia (most do not return after taking part in Wool's ambush)
2,500 Vermont Militia

24. Sir George Prevost, courtesy of John R. Lee

Victory Conditions

For the British to win, they must drive the Americans from the field. Consult your favorite rules to see how many casualties an assaulting force can suffer before breaking off their attack.

Scenario #3 – Delaying Action

The British assault also includes two smaller actions which lend themselves to be created on the tabletop. The first is Maj. Wool's ambush of the British when they were advancing on Plattsburgh. The newly arrived Crown forces, veterans of the Napoleonic Wars, viewed the Americans with contempt. The British did not put forward any light troops to screen their advance, so the Americans were able to ambush them successfully. Though the militia fled when charged by the British in column, the US regulars were able to provide an orderly retreat. A few miles later, the Americans were able to duplicate their initially success and once again extricate themselves before being overwhelmed by British numbers.

When gaming this delaying, the American player needs to focus on inflicting the largest number of casualties without being overrun. The more casualties they cause, the more victory points they should accumulate. At the same time, they cannot overstay their welcome, as the enemy wants to charge them directly. Thus, the American player needs to pick good positions that make it more difficult for the Crown forces to launch a successful charge, fire and then retire to repeat the same procedure down the road. For the British player, casualties are an inconvenience (one needs to break eggs to make an omelet), but the main goal is to sweep the Americans away and if possible, capture the guns. The guns move slowly and once across the safety of the bridge, the Americans need to spend a turn pulling up the planking. Once the planking is pulled, the game comes to an end.

For gaming purposes, though the entire British Army is present, the British player should only need to deploy one brigade of infantry, along with the squadron of dragoons, to deal with the troublesome Yankees. The more casualties taken, the more hesitant the British player should be in charging the next phase.

To increase the chance of the American troops pulling up the planking on the bridge before everyone was across, we left this up to chance. For every British unit within the line of sight of the bridge, a die is rolled at the start of every turn. Once two 6s are rolled, artillery can no longer move across the bridge. Once three 6s are rolled, American infantry can no longer cross the bridge.

When we played this scenario, the Americans were allowed to place their troops first on the map and the game began with their first fire on the British who were marching on the road. After successfully discharging two volleys, the American regulars retreated to a second prepared position, while the main body of militia ran off the table. At the second encounter, the US Rifles stayed too long and were chopped to bits. Rather than risk the capture of guns, the American player had deployed his cannon (along with a company of militia) at the third defensive position, where they did stellar service before one was captured (the other was dragged off just in time before the British dragoons arrived). As it played out, both the British and the Americans were able to claim victory.

Scenario Order of Battle

British - 3,500 men
Maj. Gen. Manley Power
3rd Regiment of Foot
5th Regiment of Foot
1st Batt., 27th Regiment of Foot
58th Regiment of Foot
Royal Artillery Battery, five 6-pdr. guns and one howitzer
19th Light Dragoons, 1 squadron

US *1,100 men*
Maj. John Wool, Twenty-Ninth US Infantry Regiment, 250 men
Maj. Appling, First US Rifles, 110 men
Brig Gen Benjamin Moorer, 700 New York State Militia
Two 6-pdr. guns

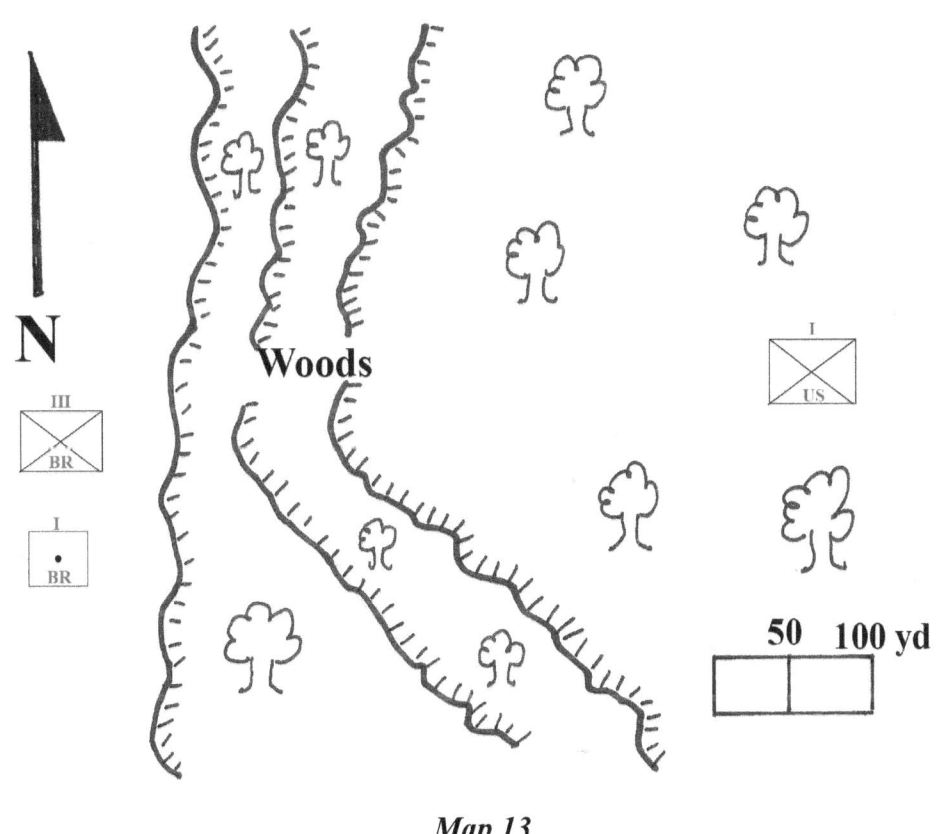

Map 13

Scenario Victory Conditions

The mission goals for the two sides are different and thus it is possible that both sides can claim victory in this scenario. However, this results in no bragging rights and what is the fun in that? To determine victory, the side with the more victory points wins.

The Americans need to cause as many casualties as possible and if possible, try to save the guns. The Americans gain 1 victory point for every British soldier they remove from the field and 5 victory points for every officer. The Americans will lose 1 victory point for every officer casualty they suffer. The British have a strong urge to charge the guns. If the forces of the Crown capture a gun they gain 20 victory points. At the end of the

25. British Squadron, Warlord Miniatures

game, the side with the most victory points wins.

Scenario #4 - Sortie on the Rocket battery

The sortie of Capt. McGlassin on the Congreve rocket battery is the stuff of 19th century legends and 20th century movies. Indeed, years after the encounter, the British officer who commanded the battery stated the assault was "the most gallant thing accomplished by any man". For those of us who watched John Wayne as Davy Crockett spike the enemy guns before the Alamo, McGlassin's assault on the Rocket Battery stands out to be recreated on the tabletop.

During the night of the 10th, a British work party was assembling the wooden "A" frame supports for their Congreve Rockets which by morning would be deployed against Ft. Brown. To deal with this threat, Capt. McGlassin, arose from his sickbed and crossed the Saranac River with fifty volunteers. With the gunflints removed from their muskets, the Americans planned to take the battery with bayonets. Unobserved in their crossing, McGlassin split his men into two groups. After one group made its way to the rear of the battery, McGlassin shouted "Charge! Men, charge! Upon the front and rear!". The British troops, unarmed except for picks and shovels, fled the battery. McGlassin's men then destroyed the battery. McGlassin returned to the American line without a single loss of life.

Any wargaming of a stealthy night assault requires the assaulting force to pass various checkpoints undetected, in this case one while crossing the river and the second one when each group approaches the battery. To keep the game flowing, in our recreation, we had made the assumptions that the Americans were able to cross the river and reach the edge of the British battery undetected. During our gaming of this event, we gave the British troops one musket for every ten men, or 30 armed men. To duplicate the actions of that evening, we divided the British into 10 separate groups, a third of whom were actually working, while the rest were supervising them from the comfort of their campfires. Thus, if the Americans are able to approach the British undetected, they need to maximize their surprise and rout as many British groups as quickly as possible. The British commanding officer, who should be placed in the center of the battery, historically tried to rally his men, but when assaulted from two sides, his men ran and he realizing he was alone, followed their example.

In assaulting the ten clusters of British troops, the first groups the American encounter should be easily defeated, but as they run across additional troops, both the shock of the attack and the adrenalin rush maybe subsiding and should be determined by a die roll. Thus, at the start of the assault, the American should be gaining significant modifiers (+1 for charge, +1 for being charged in the flank, +1 for fresh troops, +1 for being better armed), but as the battle progresses, the level of fatigue needs to be checked for each additional melee. In our wargame, we roll a die to see if the adrenalin level has worn off (a die roll of a 1, 2, or a 3 on a D6). If so, the American group first loses its +1 being fresh troops. For each following melee, the Americans would also dice to see if they have the benefit of charging the enemy in the flank. If the Americans fail the flank attack die roll when attacking one British group and are able to defeat them, they may still have the flank attack bonus when attacking the next group, so a die roll is needed.

This scenario can be played with two players, with the British player frantically trying to assemble some sort of command and control over the various groupings, but it also lends itself to solo wargaming. In either event, first the British troops are placed randomly within a 24 inch radius of their commander, as are three markers for the battery. To completely destroy the battery, all three components of the battery must be put to the torch. As this point, the Americans can launch their attack from two sides. To duplicate the stress of this assault, the game is limited to ten turns. Once the Americans defeat a unit, in the next turn they can move forward and attack the next unit (if it is within charge range) or try to destroy the battery. The American units should be strong enough to absorb a few casualties before becoming ineffective, but not too strong. If the US forces are trying to set fire to the rockets or partially completed A frame supports, the unit needs to spend an entire turn lighting the blaze. Historically the Americans defeated all of the British troops and then set fire to the battery, but this sequence of events is up to the player. Though no British troops came to the rescue, in our game we have a battalion of British troops available to counterattack the assault. Once the American player lights their first fire, the British players rolls a die (needing a 6 on a D6) to see if a British unit appears at the table's edge at the end of the turn. Once the second fire is lit a second die is rolled (needing a 5 or a 6). With only ten turns in the game, the American player needs to move quickly if they want to gain a stunning victory.

Scenario Order of Battle

Capt. George McGlassin
60 Americans

300 British

Scenario Victory Conditions

For the British to win, they need to keep the battery from being destroyed. For the Americans, a complete victory is the destruction of all three battery stores and a successful withdrawal to the American line. If the Americans destroy all three but in turn are themselves killed in the assault, they become 19th century legends, much like Gen. Custer was treated up until the 1950s (think of the 1941 movie "They Died with their Boots On"). The destruction of 1 or 2 battery stores and a successful withdrawal is a partial American victory, while if the British player manages to protect 1 or 2 of these stores and destroys one of the American units, the game is treated as a partial British win.

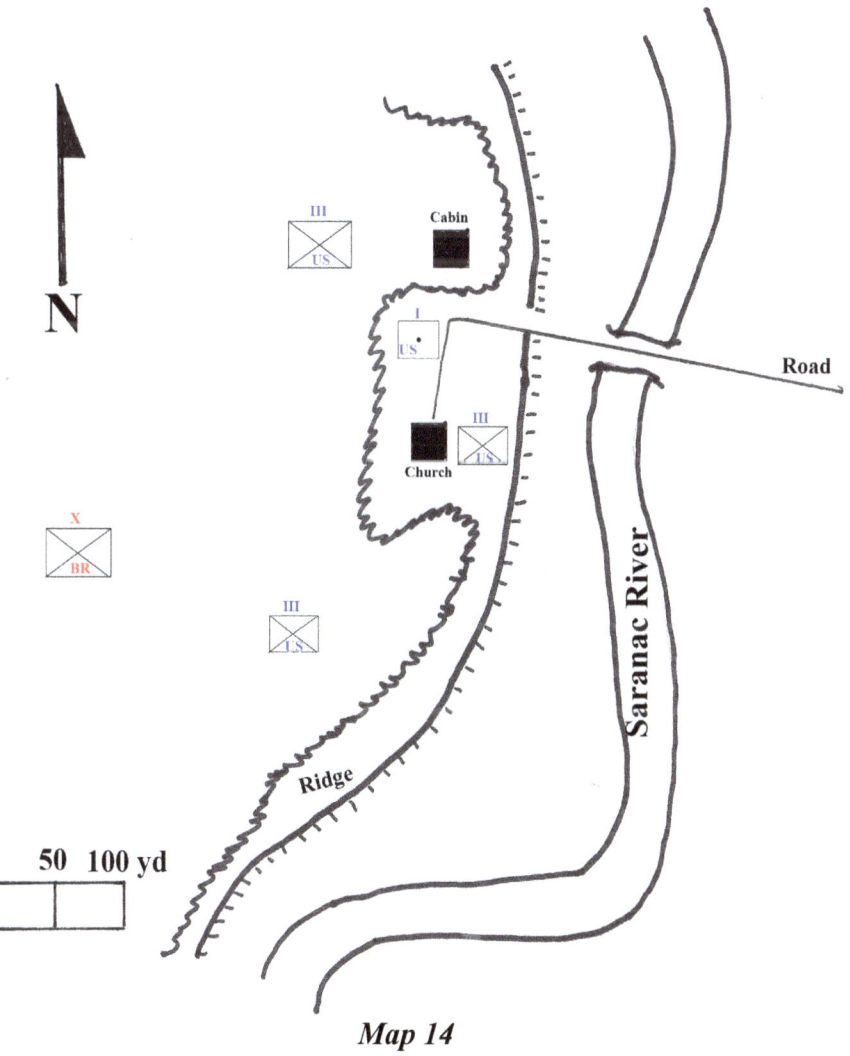

Map 14

Bad Roads and Poor Rations

26. 58th Foot, courtesy of Perry Miniatures

Further Reading
Elting, John R., *Amateurs, To Arms!: A Military History of the War of 1812.* (Da Capo Press. New York. 1995)

Everest, Allan Seymour, *The War of 1812 in the Champlain Valley.* (Syracuse University Press. Syracuse, NY. 1981)

Fitz-Enz, David G., *The Final Invasion: Plattsburgh, the War of 1812's Most Decisive Battle.* (University of Nebraska Press. Lincoln, Nebraska. 2001)

Lossing, Benson J., *Pictorial Field-Book of the War of 1812.* (Harper& Brothers Publishing. New York. 1869)

Schroeder, John H., *The Battle of Lake Champlain.* (University of Oklahoma Press. Norman, Oklahoma. 2015)

27. Rocket Battery, courtesy of Knuckleduster Miniatures

28. US troops in prepared positions, courtesy of Mike Vasile

Central Lake Ontario

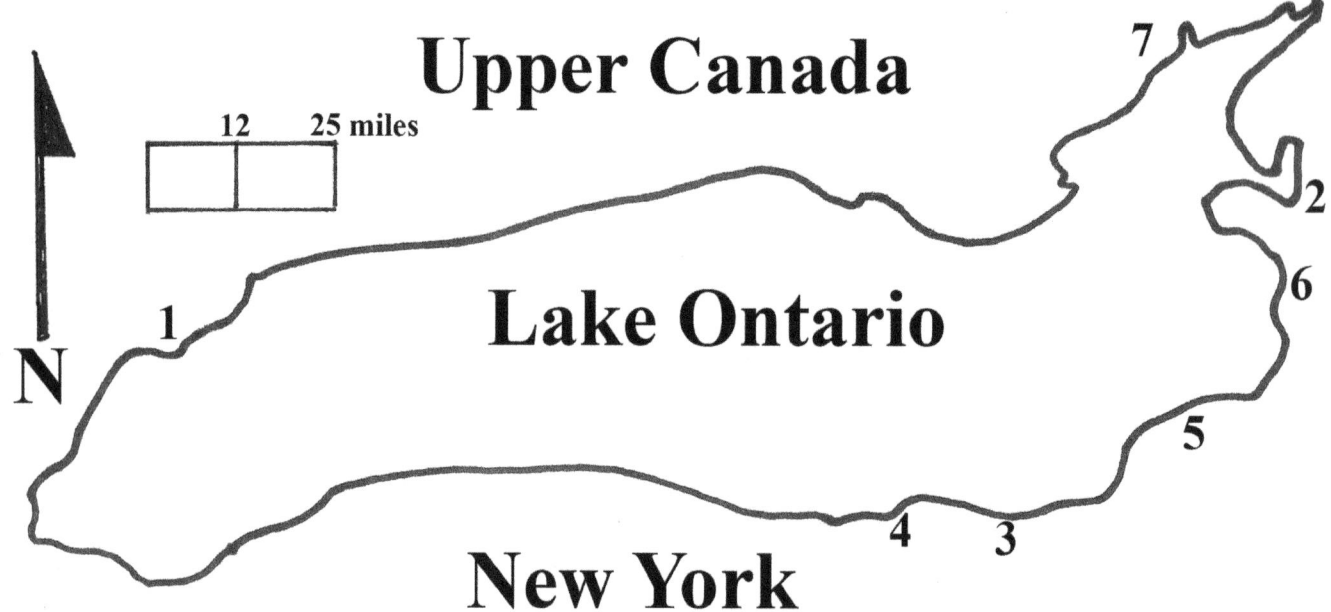

1 - York 2 - Sackett's Harbor 3 - Sodus Point 6 - Putneyville
5 - Oswego 6 - Big Sandy 7 - Kingston

Map 15

York, 27 April 1813

29. Zebulon Montgomery Pike by Charles Wilson Peale 1808, National Portrait Gallery

Overview
Large Raid - 3,500 participants

Americans force a beachhead and pushed back the outnumbered British defenders. The retreating British set fire to their supplies and blew up their gunpowder magazine. The shell-shocked troops then sack York.

Introduction

In the spring of 1813, the American Navy had achieved a temporary advantage over their British counterpart and Isaac Chauncey, the US Navy commander on Lake Ontario, was willing to engage the enemy. Believing the British naval base at Kingston was too heavily fortified, Chauncey and Dearborn pivoted from their initial target to a much less defended target – the provincial capital of York.

York, often called Little York, was rather small for a city. Some American estimates suggested that in 1813 3,000 Canadians called York home, but in reality its population was well under a 1,000. The strategic value of York was its shipyard. Boats could be repaired and new ones were being constructed, including new warships. Moreover, York was considered to be lightly defended.

Guarding the entrance to the small bay was Ft. York, which was established in 1793 by Lt. Col. John Graves Simcoe, the Lt. Governor of the newly created province of Upper Canada. This original fortification consisted of a few wooden huts surrounded by a palisade and a blockhouse. In 1811, Maj. Gen. Sir Isaac Brock was appointed the Commander in Chief for Upper Canada and he began to strengthen York's fortifications. A small irregular fort was built between Garrison Creek and the lakeshore to protect the Government House and a powder magazine was constructed. A series of earthen batteries were started but other planned building projects remained unfinished by the spring of 1813.

The weather on Lake Ontario was unpredictable and in April ice could still be encountered. Eager to begin the offensive, the American force embarked at Sackett's Harbor only to be delayed by bad weather. On 25 April the fourteen-boat flotilla departed Sackett's and arrived off of York on the 26th late in the day. Thought the delay due to the weather had allowed the British schooner *Prince Regent* to leave her winter berth at York, the nearly completed 30-gun *Sir Isaac Brock* lay in stocks and was almost ready for launch.

The Battle
On the morning of the 27th, Chauncey's goal was to land the American forces west of York near the ruins of the abandoned French fort. Strong winds, however, pushed the boats further west. As the first wave of Americans approached the wooded shoreline, the Native Americans began shooting at the boats. The riflemen returned fire as the first three companies of the Fifteenth Infantry Regiment reached the shore. Led by Gen. Pike, the three

30. Ft. York, C.W. Jeffrey

companies formed up on the beach and pushed inland. At this point, a British grenadier company had marched to the landing site and launched a bayonet charge. Though the Americans were temporarily pushed back, more troops land and they stabilized the American line. Other British troops arrived at the landing site, but they were unable to stop the landing. Faced with overwhelming odds, the British began to retire to York.

Over the next two hours, the Americans landed the rest of their force while Chauncey's squadron moved to engage the British land batteries. A misfire at the Western Battery resulted in an explosion. The remnants of the grenadier company which had retreated to the Western Battery were decimated and the company ceased to function as a unit. Though the British were able to reactivate a non-damaged gun, the overall situation was deteriorating rapidly. Gen. Sheaffe removed himself to the rear and provided no leadership at the battery as the militia faced the American advance. Unable to depress the muzzle of the remaining gun low enough to impede the coming assault, the remaining British troops fled.

At this time, Sheaffe focused on extracting what British troops he could, while abandoning the militia to their fate. With the British flag flying over the fort, Sheaffe ordered the fort's magazine blown up. As the Americans moved toward Ft. York itself, Gen. Pike stopped to question a British sergeant. At the moment, a few hundred barrels of gunpowder exploded. Gen. Pike was mortally wounded as were two of his aides, and another American officer was killed outright. The American advance stalled as Pike was carried from the field. As Sheaffe retreated past the uncompleted *Sir Isaac Brock*, he gave orders to burn the unlaunched ship and told the militia officers to ask for terms of surrender. Retreating toward Kingston, Sheaffe abandoned his two six-pdr. cannons on the muddy road east of York.

After Pike's injury, Col. Pearce of the Sixteenth U.S. Infantry Regiment assumed command and the Americans occupied York. The shell-shocked US troops noted the fires consuming the *Sir Isaac Brock* were set after the militia began their negotiations of surrender. The discovery of a human scalp by US troops in the York Parliament House did not help matters. Over the next five days, American troops looted a number of buildings, including the public library, and set fire to numerous structures when they depart on 1 May. Other military buildings, such as the barracks and the storehouse at Gibraltar Point, remained standing.

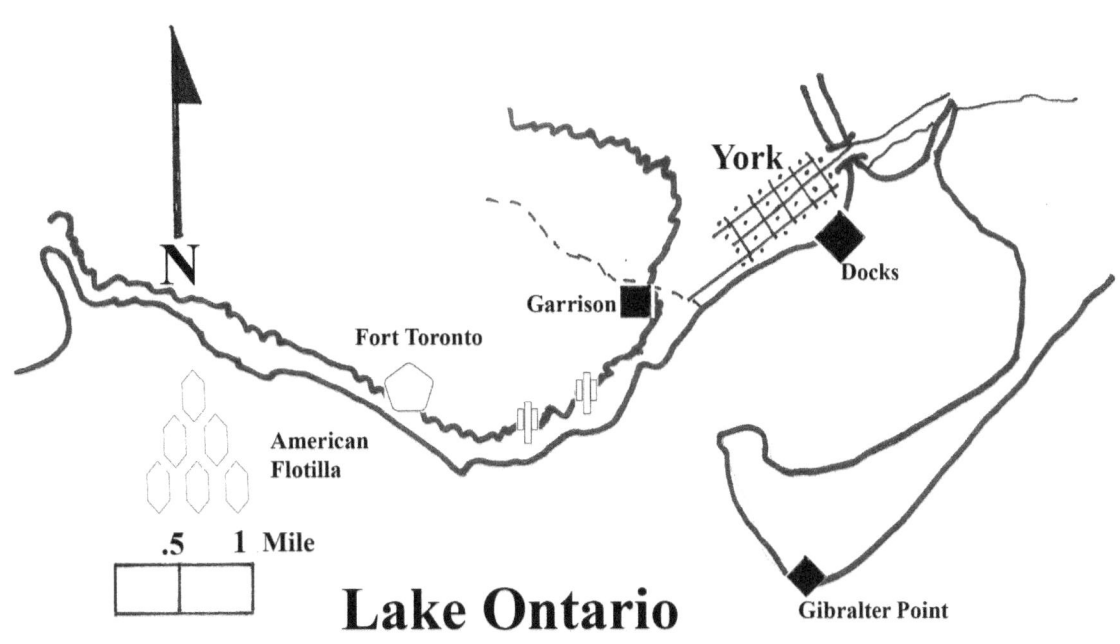

Map 16

Order of Battle

British -700 men
Commanding Officer: Gen. Sir Roger Sheaffe

Glengarry Light Infantry, 40 men
Royal Newfoundland Regiment, two companies, *120 men*
8th Regiment of Foot, two companies, *120 men*
8th Regiment of Foot, grenadier company, *60 men*
Native Americans - Mississauga and Chippewa, 80 men
Militia, *300 men*
Royal Artillery, 13 men
At the mouth of Garrison Creek
    Two 6-pdr. guns
At Government House
    Battery of two 12-pdr. guns
At the Western Battery
    2 heavy cannons (previously condemned guns without trunnions mounted to logs)

US - 1,700 men
Commanding Officers:
Maj. Gen Henry Dearborn
Commodore Isaac Chauncey
Brig. Gen. Zebulon Montgomery Pike

US Rifles, 300 men
Sixth US Infantry Regiment, *250 men*
Fourteenth US Infantry Regiment, 250 men
Fifteenth US Infantry Regiment, *350 men*
Sixteenth US Infantry Regiment, *250 men*

Twenty-First US Infantry Regiment, *250 men*
Two 6-pdr. Battery, *50 men*

President Madison, Corvette, 24 guns
Oneida, Brig, 18 guns
12 Schooners

Game Notes

The American goal, at least according to the US naval commander Commodore Chauncey, were the naval stores at York, including the uncompleted *Sir Isaac Brock*. From Chauncey's perspective, the addition of the *Brock* to the Royal Navy would be considered catastrophic for the American cause, while its incorporation into the US squadron would provide the Americans with a definitive edge. In order to avoid landing American troops in the face of established artillery positions, the Americans chose to storm a less prepared position.

The best chance the British had of stopping the Americans was when they were landing. Sheaffe correctly predicted that the Americans would land to the west of York, but as the wind shifted the actual landing site further westward, the British units were committed haphazardly. Had Sheaffe taken a more direct role in directing this initial clash, it is possible it could have been halted. In addition, had the British brought up their 6-pdr. battery to the landing, the ability of the Americans establishing a beachhead would have been more strongly contested and possibly the outcome would have been in doubt.

Once the Americans were successful in establishing a beachhead, the outnumbered British needed to delay the American advance as much as possible so that any stores could be destroyed before falling into enemy hands. Historically, what the British troops failed to accomplish in direct confrontation, they did when they detonated the powder magazine. At the point, the American assault stopped, the regular British troops withdrew, and the *Brock* was destroyed.

31. US Infantry on the march, Pendraken conversions, author's photo.

Contested landings make for interested wargames, especially when it is unclear where one side will disembark. In this recreation, the American player followed the historical plan and landed to the west of the positional batteries. As the Native Americans were patrolling the lakeshore, they are allowed to place their figures on the battlefield before any American forces were able to land.

Using the historical order of attack, seven boats filled with three companies of riflemen and the Fifteenth US Infantry approach the shore out of range of the British heavy batteries. Each boat is considered to carry a company. The first boat lands on turn 1. Each subsequent boat rolls a D6 to determine upon which turn they land. On a roll of a 1, they also arrive on turn 1. On a roll of a 2 or 3, they arrive at the start of turn two; on a roll of 4 or 5 they appear at the start of turn three. If they player rolls a 6, they arrive at the star of turn four. Once the Americans drive the British from the beach, the remaining US troops can land without incident.

Troops do not need to get out of their boats automatically and can remain in soft cover until ordered to assault the beach. Though the current and the wind pushed the Americans further west than they originally intended, for game purposes we presume that these factors effected each craft similarly. However, one can also use a pair of D6s to determine how far each boat landed from each other. To reinforce the random nature of the wind, every time the Americans land, the exact distance will need to be rerolled.

The placement and arrival of additional British troops is determined by die rolls. In our recreation, we rolled a die to see which of the following British units (one company of Glengarry Light Infantry, two companies of Royal Newfoundland Regiment, two foot companies of the 8th Regiment of Foot, the grenadier company of the 8th Regiment of Foot, three groups of Canadian militia, and the two 6-pdr. battery) arrive to stem the American assault. For each unit, a 5 or a 6 on a D6 results on the unit appearing on the tabletop. However, as each of these units may appear independently from each other, a second D6 die roll determines which turn they arrive. Arriving units always appear at the table edge at the end of the turn. This recreation is set to run for ten turns.

If one were to take a broader approach in wargaming the assault, one can try to force a landing closer to York itself. In one recreation recently played, the Americans landed to the east of York. Though the landing was opposed by Native Americans and British regulars, the Americans were able to establish a beachhead and move inland while the US Navy continued to bombard the British land batteries. In this game Sheaffe died heroically by the two 6-pdr. battery. Gen. Pike was then able to secure the *Brock* and his troops were able to extinguish two small fires lit by the British before they did extensive damage. In another attempt, the American player chose to land in the face of the British guns and with some good die rolling, the British player was able to sink six of the seven landing craft. With the first wave of the invasion defeated, the Americans never launched a second wave.

Victory Conditions

The American goal is to land in the face of the enemy and capture or destroy the British public works. Consult your favorite rules as to how many casualties they can suffer before their force has to withdraw. The game plays for ten turns.

Further Reading

Elting, John R., *Amateurs, To Arms!* (Da Capo Press, Inc. NY. 1995)
Lossing, Benson J., *Pictorial Field-Book of the War of 1812.* (Harper & Brothers Publishing. New York. 1869)
Stacey, C. P., Col., *The Battle of Little York.* (The Toronto Historical Board. Toronto. 1977)

32. Royal Newfoundland Fencibles, courtesy of Perry Miniatures.

Sackett's Harbor, 29 May 1813

Overview

Small Battle – 3,500 men

British raid to destroy the principle American naval facilities on Lake Ontario

Introduction

Sackett's Harbor was the principal US shipyard on Lake Ontario. The goal of the Crown forces was to destroy the naval stores at the shipyard, including the soon to be completed *USS General Pike*. When the British learned that most of the US garrison was away from Sackett's, Lt. Gen. Sir George Prevost, the Governor of Canada, quickly put together a force of six ships and almost 1,000 men, including 40 natives. As Sackett's was raided the previous year, the Americans had reinforced the base with more cannon, abatis (defensive works made from fallen trees), earthworks and blockhouses. A small number of regular troops were stationed at the two forts at the harbor entrance, Fort Volunteer and Fort Tompkins. In addition, the defenders included detachments of mounted and dismounted dragoons, invalid troops, seamen, US Marines, trained volunteers and militia. All American forces at Sackett's were under the command of New York State Militia Brig. Gen. Jacob Brown.

The Battle

An American schooner spotted the British squadron the previous evening and was able to warn Gen. Brown of the coming attack. Rather than sail directly into the harbor, Prevost landed his men south of the base at Horse Island. The landing was opposed by a combination of artillery fire from two 6-pdrs., and long-range shelling by a 32-pdr. firing from Fort Tompkins and the Albany Volunteers. The British gunboats returned fire and killed Col. John Mills, the commander of the Albany Volunteers. As the British pushed forward, the Americans fell back and the British successfully crossed the flooded causeway which linked the island and the mainland. Commodore Yeo was at the head of this advance and urged his men forward by waiving his hat. The overall commander of the raiding force was Col. Edward Baynes, who split his force into two equal columns. One group, made up of regular British troops and the Newfoundland Fencibles, pushed along the shoreline towards Fort Tompkins. The second group, consisting of Light Infantry, Canadian Voltigeurs, and Native Americans, advanced through the woods. A running battle took place in the woods between causeway and the town itself, with individual companies fighting among the abatis. The US forces continued to retreat to the earthworks and Fort Thompkins closer to the harbor. The British tried to press their attacks, but the lack of British naval artillery support failed to dislodge the Americans from their defenses. One British boat, the *HMS General Beresford* was able to get close enough to Fort Thompkins to silence its 32-pdr. The now rallied American Militia continued to pepper the Redcoats along their exposed flank. Fearful of being cut off from his landing craft by the large numbers of militia within the woods, Prevost successfully extracted his force and sailed back to Canada. During the course of the raid, miscommunication and possible panic amongst the naval personal resulted in the US Navy setting fires to their own stores and the frigate *General Pike*.

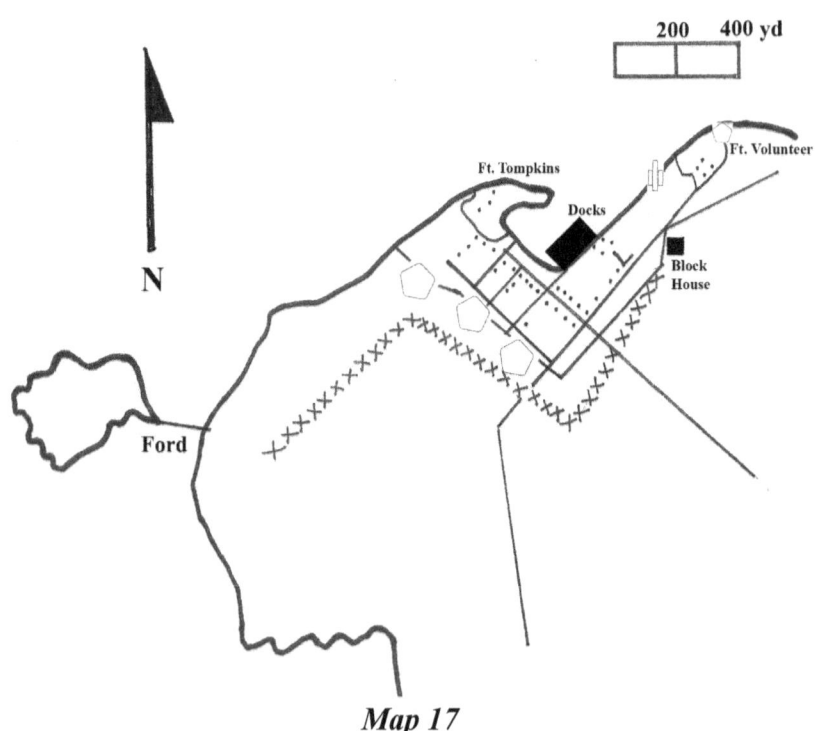

Map 17

Order of Battle
British - 1,000 men

Commanding Officers:
Lt. Gen. Sir George Prevost, Senior Commander

Col. Edward Baynes, Commander of Landing Force
Commodore Sir James Lucas Yeo, Commander of Naval Force, accompanied the landing force

1st (Royal Scots) Foot, 1 section, 25 men
8th (Kings) Foot, 2 companies, 200 men
100th Regiment of Foot, grenadier company, 56 men
104th Regiment of Foot, 4 companies, 330 men
Royal Newfoundland Fencibles, *1 company, 60 men*
Glengarry Light Infantry Fencibles, 1 company, 46 men
Canadian Voltigeurs, 2 companies, 120 men
Royal Artillery, 2 6-pdrs. and 80 men, never disembarked
Native Americans, 40 men

33. 3rd US artilleryman armed with a musket, courtesy of W Britain.

British Naval Craft
HMS Wolfe   Ship, 18x32 carronades & three 18-pdr.
HMS Royal George   Ship, 18x32 carronades, three 18-pdrs., one 24-pdr.
HMS Earl of Moira Brig, twelve 24-pdr. carronades, one 18-pdr.
HMS General Beresford
Sir Sydney Smith   Schooner, 10x32 carronades and two 12-pdrs.
Lady Murray
2 gunboats
30 Bateaux
Canoes

US - 2,500 men
Commanding Officers:
New York State Militia Brig. Gen. Jacob Brown
Lt. Col. Electus Backus, US Light Dragoons

First US Light Dragoons, *200 men*, dismounted
First US Light Dragoons, *120 men*, mounted
1st US Artillery, *100 men*
3rd US Artillery, 134 men, armed with muskets and fought as infantry
Heavy Artillery   men, 8 men
Ninth/Twenty-First Infantry Invalids, 315 men
Twenty-Third Infantry, 1 company, 34 men

Col. John Mills
Albany Volunteers, 250 men
Two brass 6-pdrs., *60 men*

New York State Militia, *650 men*
Fifty-Fifth New York State Militia, *350 men*
Seventy-Sixth New York State Militia, *300 men*

US Navy
USS General Pike (under construction)	Frigate, 26 24-pdrs.
USS Fair American	schooner, one 32-pdr. and one 12-pdr.
USS Pert	schooner, one 32-pdr. and two 6-pdrs.
USS Duke of Gloucester	recently captured and under repair

US Marines, 1 company - not engaged as they were posted at or near Fort Pike

Game Notes

Sackett's Harbor is a fun scenario to recreate on the tabletop. At its essence, the game pits a small core of professional British troops against a larger hodge-podge of raw recruits and invalids (shoot well but move slowly). The Americans have three major opportunities to stop the monarchists – 1) during the initial landings, 2) the open causeway between Horse Island and the mainland, and 3) the defensive line. The American player should also consider a defense in depth – like Cowpens. Conversely, the forces of the Crown should identify their primary objective and commit to the plan.

The British initial advantage in troop quality should decrease as the game progresses, as this reflects the heavy casualties suffered by the Crown forces (approximately 25% of the British invasion force was killed, wounded or missing). Both Prevost and Brown commanded from the front, so they can influence particular units, but cannot impact events outside their field of vision. The British never used the brass 6-pdr. that they captured from the American militia, since their landed force did not include any gunners. If you do choose to allow the captured gun to be used, I would suggest you reducing its rate and accuracy, as well as not being able to drag it into the woods. While within the abatis and the woods, troops need to operate in open order, with each company acting independently.

Both sides have strengths and weaknesses and more importantly, have an equal chance of winning. The British goal for this attack is the destruction of the US naval stores and the possible capture of former British ships now in US service. The American goal is to drive off the British and protect the naval stores. Historically, the Americans were able to repulse the British land forces, but a miscommunication/panic resulted in the Americans setting fire to their own stores. Thus, the British were unable to defeat the American forces but came rather close to achieving their more important goal of destroying the naval supplies and thus gaining naval superiority.

Any recreation of the battle on the tabletop needs to replicate the American eagerness of not allowing the naval supplies to fall into enemy hands. In our recreation of this event, when a British unit move within six inches of either Ft. Tompkins or Ft. Virginia, a die roll is required to

34. Author's Sacketts Harbor game, circa 1996

see if either the supplies or the *General Pike* is set afire. This process is repeated for every subsequent British unit that appears in this area.

Each rule set may have a mechanism of setting fire, but we have used this one over the years. Once a British unit is close to its objectives, it is possible that over eager US personal will set fire to their own supplies to keep them out of enemy hands. On a roll of a 6 on a D6, a fire is set and a second die role of a D6 identifies what is burning – a result of a 1, 2, or 3 indicates naval stores on land while a result of a 4, 5, or 6 indicates that it's the warship. To fight a blaze, a US unit is needed to try to stop the flame. Each US objective is considered to have one unit available to fight the fire, but they must be ordered to do so. A senior officer needs to be within 12 inches of the burning item to give the order to put out the fire. If additional units are present within 6 inches of the blaze, they can also try to stop the fire.

To put out a fire, a roll of a 6 on a D6 is required at the start of the turn for each unit fighting the blaze. Every turn the fire is not put out, the harder it is to extinguish. For every turn the fire continues to burn, an additional fire marker is added to the objective. The fire will continue to burn for six turns. At the end of the sixth turn, the blaze will have completely destroyed the stores or burned the ship to the waterline.

Victory Conditions

The British need to destroy the American naval facility before they suffer too many casualties. By all means, consult your favorite rules as to how many men the British can lose before withdrawing, but given Prevost's concern for the loss of life (either on humanitarian grounds or knowing how difficult it would be get replacements), the number should not exceed 20% percent. The Americans win if they stop the British from reaching and setting fire to the naval facilities.

Further Reading

Hough, Franklin, *A History of Jefferson County in the State of New York.* (Joel Munsell. Albany, NY. 1854)

Lossing, Benson J., *Pictorial Field-Book of the War of 1812.* (Harper& Brothers Publishing. New York. 1869)

Wilder, Patrick A., *The Battle of Sackett's Harbour: 1813.* (The Nautical and Aviation Publishing Company of America. Baltimore, MD. 1994)

35. US Light Dragoons (dismounted), courtessy of Richard Watts.

Sodus Point, 29 June 1813

Overview
Small Skirmish - less than 500 participants

British raid on the American public stores at Sodus Point. The initial British landing is repulsed but the British troops return the following day and burn the village to the ground.

Introduction

During the first year of the war, when the British forces raided American settlements on the southern shoreline of Lake Ontario, the encounters were quite cordial. Even after the rampage and destruction of both public and private property at York by the shell-shocked American troops, a level of civility continued to be observed. When the British captured the Port of Charlotte on 15 June 1813, for example, they gave a receipt to the clerk for supplies they took.

Yet attitudes were beginning to change and even those American residence who were lukewarm to the war, began to take an active role in the defense. British troops who had previously fought against Napoleon and were being sent to North America also perceived the enemy in less chivalrous terms. This change in attitude toward the treatment of civilians will become even more pronounced during the following year when the British begin to raid the Chesapeake Bay area at will.

The village of Sodus, often referred to as Big Sodus or Sodus Point, was first settled in the last decade of the 18th century. By the start of the war, two mills – a gristmill and a sawmill, a tavern, a warehouse and several homes constituted the settlement. Recognizing the value of the goods held in Sodus, the local population participated in a militia and continued to be vigilant of any sighting of a British flotilla. When news of the British appearing at Charlotte filtered to Sodus, Col Swift's regiment of militia mustered at Sodus, but after a few days with no signs of British activity, dispersed after removing all public stores inland.

36. British landing, courtesy of Perry Miniatures.

The Battle

Shortly after the militia regiment began to disburse, the British flotilla appeared at Sodus in the early evening. Perhaps not wanting to lose the element of surprise, the British sent in sixty men of the 1st battalion of Royal Scots, who were serving as Marines, and a few sailors to capture the town.

Split into a small advanced party and the main body, the British approached the town as it began to rain. Besides finding a drunken man in the tavern, the town appeared to be deserted. The advanced party was then fired upon by a group of militiamen. The main body, hearing the sounds of battle, ran toward the noise. The advance party saw the body of formed men moving towards them and fired upon their countrymen. A flash of lighting illuminated the forces and brought an end to the friendly fire incident. Clearly having lost the element of surprise, the British retired to their flotilla. The American militia also retired from the battlefield after delivering their first volley.

The following morning, the British returned to the deserted village and began to loot it. The Americans had successfully hid 8,000 barrels of flour in the nearby woods and the returning British were only able to seize between 200 and 500 barrels of flour, a few barrels of pork and whiskey, and as much private property they could find. Then, as a warning against all those who would resist the British Crown, Commander Yeo ordered the town to be put to the torch. All buildings in the town, save for the tavern, were then burned. The British then returned to their flotilla and sailed away.

Impact of the Battle

Though one can debate the decision of Yeo, the Commander and Chief on the Lakes of Canada, to put the town to the torch, his actions also alarmed British citizens in Canada. Unfortunately, this contemporary concern was valid, and the precedent of destroyed civilian homes continued with the Americans burning the village of Newark in December 1813. Though the American commander was dismissed for his actions, the British burned the village of Black Rock in direct retaliation. Thus, when Washington, D.C. was captured in the following August, torches were used in its destruction.

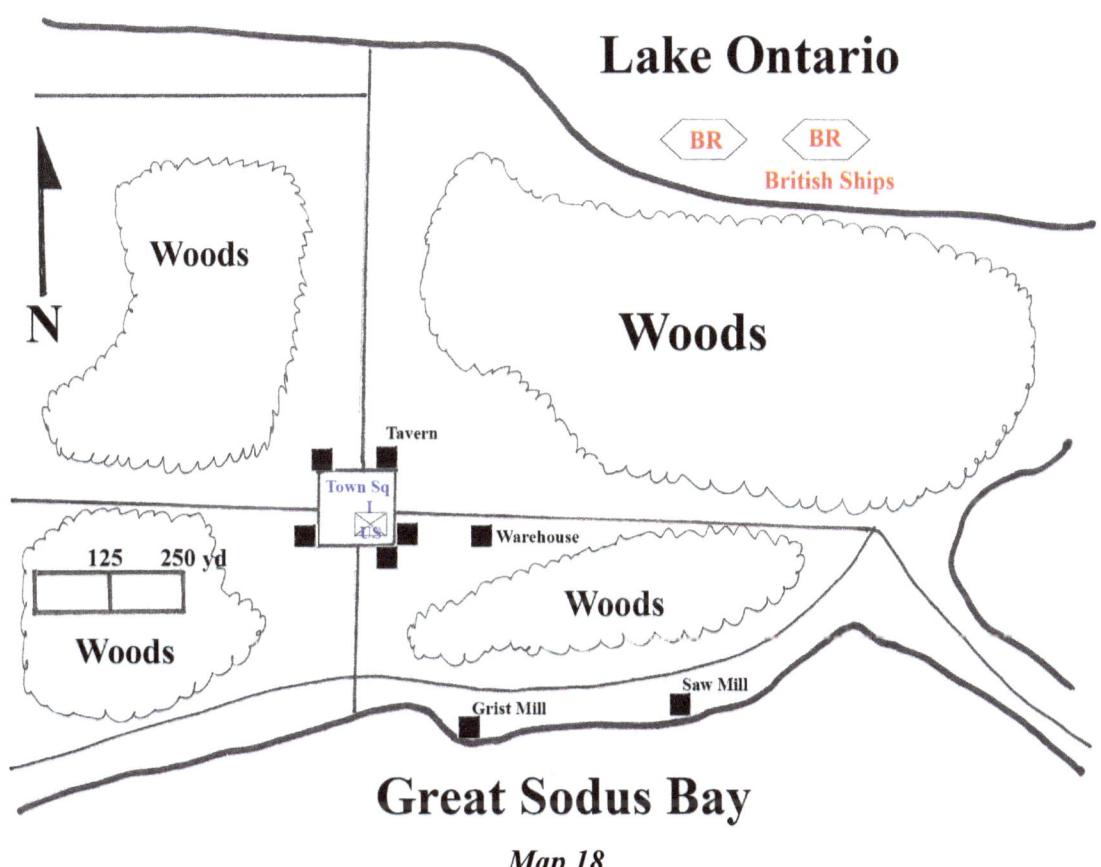

Map 18

Order of Battle
British, 100 men
Advanced party
Commanding Officer: Capt. Mulcaster
James Richardson, RN
Royal Marines, 12 men

Main body
Capt. John Wilson, 1st Battalion of Royal Scots

Royals Scots, 60 men
Royal Navy, 24 seamen

US, *60 men*
Commanding Officer: Capt. Elias Hull

Game Notes

The British raid on Sodus was unique in that it was conducted at night. Nighttime raids were always dangerous and launching one during a rainstorm could be potentially disastrous. As the British hoped to catch the Americans unaware, once they appeared offshore, they were committed to landing. Despite the inclement weather, the British were able to land successfully.

To duplicate the potential of landing troops at night on a hostile shore, the British player divides their force into five groups to represent the boats used in the landings. A D6 is then rolled for each boat to see how scattered the boats would land from their intended landing site. Divide the southern part of the table into four equal parts. For each boat, roll a D6. On a roll of a 1, 2, 3, or 4 each craft lands at the respective zone. On a roll of a 5, the defender determines where the boat will land. On a 6, the attacker choses where the boat lands.

As the landing occurs at night, any troops that land more than one movement turn from the landing site are out of visible range. To avoid giving away their position, it was common practice for troops to assault with loaded muskets and the flints removed. At Sodus, we know that the British troops fired on each other, so sometime in their assault they stopped to load, or they landed with already loaded muskets. For game purposes, the British player can land with either loaded or unloaded muskets.

Since the battle occurred in the rain, there is a high likelihood that any troops outside would be firing at a reduced rate. Personal experience of firing black powder flintlocks has demonstrated the uncertainty of the ignition system as half to a third of the shots fail to go downrange in high humidity. In rain, moisture and water would accumulate in the pan and the likelihood of a successful ignition would be further reduced.

If the British land with unloaded muskets, they must spend a turn loading before being able to fire. If the British land with loaded muskets, only half of troops would be able to fire. Any American or British troops outside

37. Local Militia, courtesy of Tod Kershner

a building may fire, but only one out of every two figures would be able to shoot in the rain. If the American player disburses their troops into buildings, they do not suffer a shooting penalty, but would be considered to be out of command. Being out of command in this scenario would allow troops to shoot, reload, and melee if attacked, but they would not be able to charge the enemy or leave the building unless ordered by an American officer.

Friendly fire and distinguishing friend from foe are always a consideration in war but it is difficult to replicate on a tabletop. In our game, any unit within the maximum distance for point blank fire[3] is considered to be unidentifiable and will be shot at by a friend. If two or more units from the same side are within musket range but are more than the maximum distance for point blank fire from each other, they automatically MUST shoot at each other for one turn. The actual damage the troops will suffer, however, is reduced by the fact that the troops are shooting in the dark and the rain; thus, only one out of every two figures are able to shoot.

The British goal of the raid was to capture supplies. Upon learning that the warehouse was mostly empty, the forces of the Crown begin to loot private houses. If a British unit is looting a building, it cannot defend itself and only half of the figures are able to engage in melee. Once a British unit suffers 20% casualties, the element of surprise is lost, and the effected unit will withdraw to the flotilla.

Ten buildings are used to represent the village and twenty supply tokens (seventeen barrels of wheat and pork, three barrels of whiskey) are secretly placed by the American player throughout the town. Each building can hold three supply tokens. Though the Americans can drive the British off, they will return the following day and torch the town. However, the more British killed, the more victory points the Americans accumulate. For the British, the deaths of the Americans are irrelevant and what matters are the amount of supplies they can gather.

When the British player liberates a barrel of whiskey, there is a chance that the troops will indulge. Once a barrel of the devil's drink is found, the unit rolls a D6. On a roll of a 1, 2, or 3, the unit functions normally, but on a roll of a 4 or 5, the unit charges the next building in search of more loot, even if under fire. On a roll of a 6, the unit no longer does anything, as the troops are too drunk to move for the rest of the game.

38. Royal Marines, author's photo.

---

[3] Each rule set uses a different number for the maximum distance for point blank fire, so please refer to your rule set of choice. If your rule set does not include point blank fire, use the regular fire range but reduce the number by half. In *All The King's Men*, for example, infantry can fire normally up to 12 inches, so for this scenario they will shoot and be shot at when within 6 inches of a friendly unit.

Victory Conditions

The American player needs to drive off the British nighttime attack before losing 50% of their force. The British player needs to gather as many supplies as possible before withdrawing. Each gathered supply is worth one victory point to the British and one British casualty is worth one victory point to the Americans. British officers are worth two points, as are barrels of whiskey.

Further Reading

Clark, Lewis, *Military History of Wayne County*. (Truair, Smith and Bruce. Sodus, NY. 1883)

Malcomson, Robert (editor) *Sailors of 1812: Memoirs and Letters of Naval Officers on Lake Ontario.* (Old Fort Niagara Association. Youngstown, NY. 1997)

Sims, Doris M., *The Battle of Sodus Point: War of 1812*. (Wayne County Historical Society. Lyons, NY. 1985)

Pultneyville, 15 May 1814

Adrian O. Mandzy

Overview
Large Skirmish - more than 500 participants

British troops raid a small settlement.

Introduction
Located where the mouth of Salmon Creek flows into Lake Ontario, Pultneyville began as a fur trading post. Incorporated in 1806, the shipping port continued to prosper and by 1820, it was home to several schooners. As in many such lakeshore settlements, the US government stored supplies and foodstuffs. Located on the way from Sackett's Harbor to the Niagara Theater of operations, American troops often stopped at Pultneyville. A public storehouse stood on the edge of the lake.

For the first two years of the war, Pultneyville was small enough to avoid the attentions of the British. In light of the ongoing conflict, the local citizens had come together for a common defense. A mounted militia unit patrolled the area and in Pultneyville itself, a militia company was formed.

Following the British burning of Sodus Point the previous year, the local citizens continued to prepare for another attack. On 14 May, when the British squadron threatened the neighboring port of Charlotte, additional militia troops were dispatched to Pultneyville. Public goods, consisting predominantly of flour, were removed from the public warehouse and 100 barrels of moldy flour were put in its place.

The Battle
On the morning of the 15th, the American militia were drilling on the streets. A heavy morning fog had blanketed the area and had obscured the British squadron. As the fog lifted, the Americans were stunned to see the British squadron at anchor outside their village. The British fired a broadside at the settlement and the militia quickly took cover in the nearby woods. The remaining town's citizens quickly followed them into the trees.

The British approached the settlement under a flag of truce demanding that public goods be turned over to them. If the Americans complied, the British promised they would respect private property. The outnumbered Americans, having previously replaced the usable war material with moldy flour, agreed to these terms.

The British force landed and quickly seized the storehouse. Unsatisfied with their moldy flour, the British began to loot private homes.

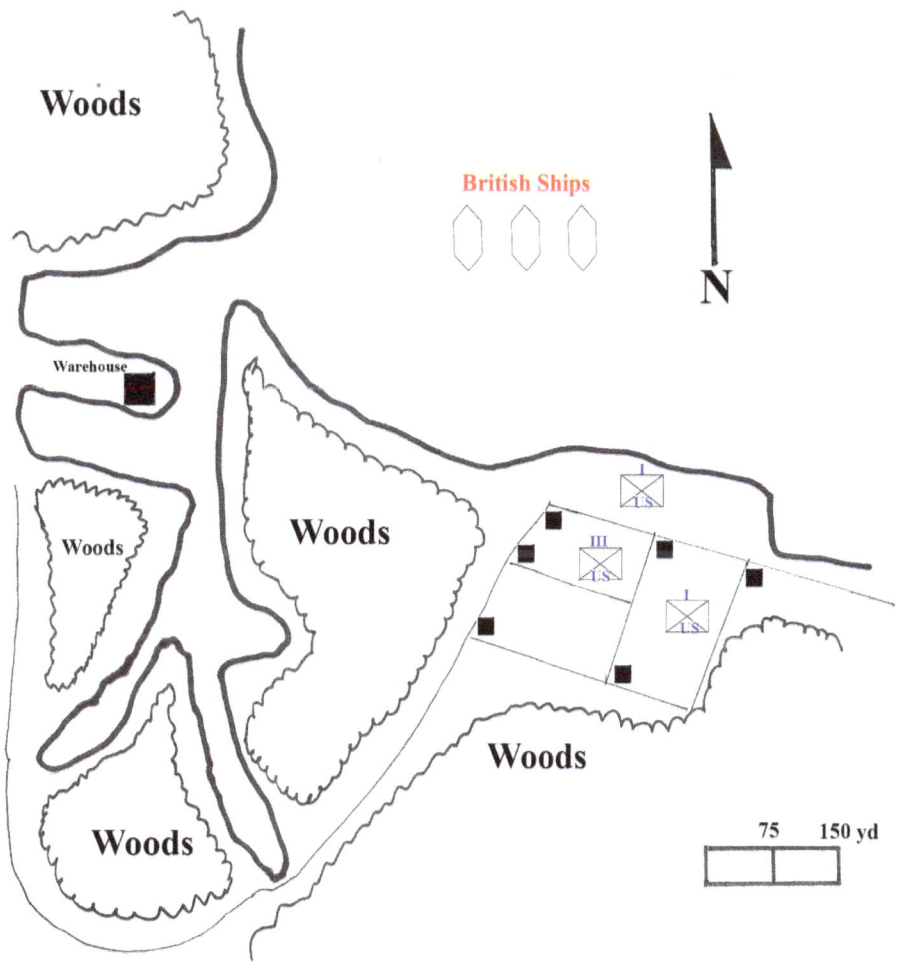

Map 19

At this time, the American militia began to fire on the British from the wood line. The British continue to pilfer and recognizing that no major stores can be found nearby, withdrew to their boats. The British squadron continued to fire on the town, but besides a few broken windows, no American casualties were reported.

39. Sailors Skirmishing with Muskets, courtessy of Perry Miniatures.

Order of Battle
British - 460 men
Commanding Officer: Capt. Short
Royal Marines, 260 men in four units
Royal Navy, 200 men in four units

US - *200 men*
Commanding Officer: Gen. Swift
Trained militia, *130 men*

Maj. William Rogers
Volunteer Riflemen, *1 company, 30 men*
Roger's Battalion, *group of 40 men,* to call them a unit suggests some sort of organization.

Game Notes

    On a first reading, the Americans have little in the way of stopping the British. The British have battle hardened troops and they have more of them. More importantly, the British have naval artillery support.

    The Americans have no hope of stopping the British advance in the town, but they do have advantages. More American troops are in the area and the British do not know how far away they are. Once the British move away from the shore, they lose their artillery advantage. The more the Americans can draw them into the interior, and into the woods themselves, the more isolated the Royal Marines and sailors become. American Riflemen, such as Thomas Fuller, are in the area and participate in the Niagara campaign later that year. For game purposes, Gen. Swifts Militia are split into two units.

    To keep the game moving, the British have ten turns to complete their mission of gathering supplies. Half of the British troops are needed to move the stores from the warehouse to the British squadron waiting off the table – which include the seamen and one unit of Royal Marines. When the British learned that 300 additional barrels of flour were recently removed from the warehouse, they detach troops to find them. In our game, we allow three units of Royal Marines and one of Sailors to find them. There are a number of houses in which the barrels may reside, and the lure of additional booty is hard to resist.

    In our game, we have a building representing a warehouse and seven other structures scattered around the table. The American player then secretly places five barrels (up to three in each house) on the table. When a British unit moves into a house to search it, they will find any hidden barrels automatically, but they roll a D6. On a roll of a 5 or a 6, the British troops begin looting for the turn. The British troops will continue to loot until they stop. To get the British troops to reform, a roll of a 4, 5, or 6 on a D6 is required at the start of a turn. If the British troops are charged when looting they fight as disordered.

For the British to win the scenario, they need to collect additional barrels of flour while suffering few casualties. For every casualty they suffer, they need to see if they continue to follow orders or if they move to attack the Americans. A roll of a 6 on a D6 indicates the British will move toward the Americans. The British will continue to engage the militia until the unit is destroyed.

Starting on turn 5, possible American reinforcements could arrive on the south, east or west roads. On turn 5 and for every turn after, a die roll of a 6 on a D6 indicates the arrival of troops. An additional die roll dictates which road is used and a third die roll states what troops are arriving – a roll of a 1 indicates company of US regulars; a 2 – a company of US Riflemen; a 3, 4 or 5 – a company of militia; and a 6 - a troop of cavalry.

40. New York State Militia Officer, courtesy of W Britain.

Victory Conditions

Each barrel of flour is worth 5 points. Each British Marine and Sailor is worth 1 point, each militia unit is worth 5 points, and each officer is worth 10 points. For the British to win the scenario, they need to collect additional barrels of flour while suffering few casualties. For the US player to win, they need to cause as many British casualties and not be completely destroyed while protecting the town. If three American units are destroyed, the Americans cannot win. Conversely, if the British fail to gain the 5 barrels of flour and get them back to their boats in 10 turns, they also cannot win.

Further Reading

Clark, Lewis, *Military History of Wayne County*. (Truair, Smith and Bruce. Sodus, NY. 1883)

Wilder, Patrick, *Seaways Trail Guidebook to the War of 1812*. (Seaways Trail, Inc. Oswego, NY. 1987)

41. Royal Navy Landing Boat with Crew, courtessy of Britannia Miniatures.

Oswego, 5-6 May 1814

Overview
Small Battle - 1,500 participants

British troops attack and capture Ft. Ontario at Oswego in an attempt to stop war materials reaching Sackett's Harbor.

Introduction
During the War of 1812, Oswego served as the central transit point for supplies being moved from New York City to Sackett's Harbor, the central US naval shipyard on Lake Ontario. In the ongoing naval arms race, Americans were close to completing a new frigate, the *Superior*, which when launched would give them a distinct advantage and temporary superiority.

Built in 1759 by the British to protect the town of Oswego, Ft. Ontario had been abandoned before the start of the War of 1812. In the summer of 1813, British sought to attack Oswego, but observing from their ships the large numbers of American troops and newly constructed defenses, they moved on and sacked the village of Sodus instead.

In early spring of the following year, Commodore Sir James Lucas Yeo believed that there were thirty plus cannons currently being stored at Oswego. Learning that Oswego was poorly defended, Yeo assembled a large force and launched an attack to capture these supplies. Though the Americans had partially rebuilt Ft. Ontario, it was not properly manned or armed at the time of attack. In May 1814, only five condemned cannons – one 9-pdr. gun and four 5-pdr. cannons – were available to the defenders of the fort.

The Battle
Learning that the British were planning to attack Oswego, Lt. Col. George Mitchell and a battalion of the Third artillery, armed as infantry, were sent to defend the supply base from Sackett's Harbor. To block the British from moving up the Oswego River, the recently captured schooner *Growler* was sunk and its crew was absorbed into the defending force. Militia was called up from the surrounding countryside and the Americans pitched tents along the shores of Lake Ontario to make the size of their force appear larger than it was.

On 5 May, the British force arrived at Oswego and attempted to land. The guns which were placed in a battery near the shore opened fire on the British. According to the Americans, the fire was devastating and a number of men were killed or wounded in the landing craft. British sources state that the winds changed and the British landing craft withdrew.

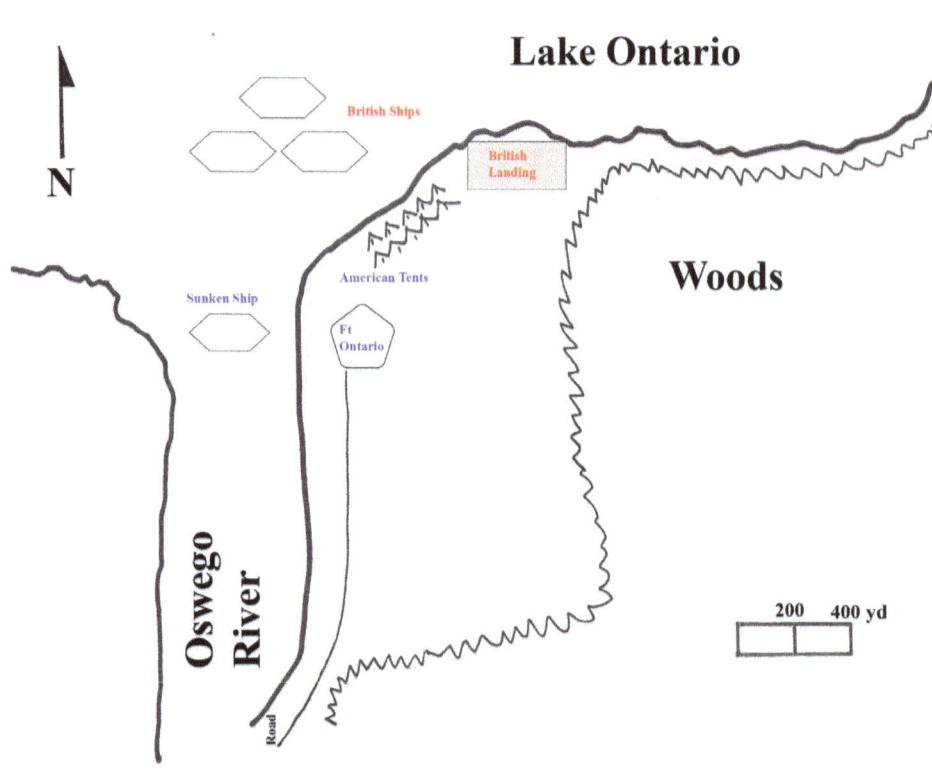

Map 20

The following day, the Crown forces returned. After a preliminary bombardment, the British forces landed three distinct groups – Royal Marines, seamen, and light troops. The light troops landed toward the east of the fort and engaged the American militia and musket armed artillerymen along the fort's eastern flank. The seamen and Royal Marines landed to the west of the first wave and made their way toward the partially rebuilt fort. Lt. Col. Mitchell had nailed the US flag to the fort's main flagstaff as a sign of no surrender. The defenders were unable to stop the British assault at either the fort or in the woods to the east of the fort and began to fall back. Having captured the fort and a number of supplies, the British proceeded to burn the fort. Though the fort was lost, the Americans claimed victory as they were able to successfully disburse a number of naval pieces that were critical to the shipbuilding program and then bring them to Sackett's harbor.

42. Fort Ontario, author's photo.

Order of Battle
British - 1,000 men
Commanding Officer: Lt. Gen. Gordon Drummond
2nd Battalion Royal Marines, 1 battalion, 350 men
Sailors, 1 detachment, 200 men
De Watteville's Regiment, Lt Col Karl Victor Fischer, 2 companies, *400 men*
Glengarry Light Infantry Fencibles, 1 company, 50 men

US - 500 men
Commanding Officer: Lt. Col. George Mitchell
Third Artillery Regiment, 242 men - with muskets or acting as artillerymen
Seamen (from the *Growler*), 25 men
Militia, 200 men
Artillery – six guns
    one 9-pdr. cannon
    four Brass 6-pdr. cannon
    one 12-pdr.

Game Notes
The attack and destruction of Ft. Ontario makes a great wargame, despite the fact the Americans cannot stop the British from capturing the fort. However, the protection of the fort is irrelevant to the war effort – it is the supplies that are critical and need to be protected. For every turn the British are stopped from entering the fort and

Bad Roads and Poor Rations

43. Ft Oswego 1814 Garrison Flag, Drummond family Castle, courtesy of Paul Lear.

the harbor below, more supplies are moved to a safe location off board. To move more supplies further or faster, the American player may remove from the fight militia troops but all of the regular troops **must** remain to fight the British (a point of honor).

The American commander is a brave and effective leader but cannot be in all places. Rivalry between the British forces as to who will capture the fort and the large US flag can be duplicated if three different players command the British forces. One player would command the Royal Marines, another the De Watteville's Regiment, and the third the sailors and Glengarry Light Infantry. Once the American forces are deployed, they did not really move and can be gamed by one player or even simply having them on an automatic response – once a British unit comes within range, they fire. At the time when the US forces cannot hold a position, they will retreat automatically.

Victory Conditions

If this attack is gamed with three players taking command of a British force, on the force that capture the American flag wins the game. If the game is played in a more traditional manner, with one player commanding the British and the other the Americans, the forces of the Crown win a great victory if they are able to capture the flag in six turns or less. If the British are unable to capture the flag at the end of turn eight, the Americans win the game.

Further Reading

Lossing, Benson J., *Pictorial Field-Book of the War of 1812*. (Harper & Brothers Publishing. New York. 1869)
Slosek, Anthony, *Oswego and the War of 1812*. Heritage Foundation of Oswego. Oswego, NY. 1989)

44. De Watteville's Regiment, courtesy of W Britain.

Big Sandy, 28 May 1814

Adrian O. Mandzy

Overview
Small Skirmish - less than 500 participants

After the Royal Navy captures a single US supply ship, the US Riflemen and Native Americans set a trap and capture the pursuing British naval force.

Introduction
Since the war began, both parties sought to dominate Lake Ontario by constructing larger and more powerful navies. To do so, both sides were engaged in a logistical race to bring the needed people and material to their respected forward areas – Kingston for the British and Sackett's Harbor for the Americans. For the Americans to supply Sackett's, supplies were shipped on the Oneida Carry, a supply chain used since the middle of the 18th century. This systems of portages and natural bodies of water entered Lake Ontario at Oswego.

The British made numerous attempts to disrupt the American fleet building program, including attacks of Oswego and Sackett's, but were unsuccessful in causing any long-term damage. Following the British destruction of Ft. Ontario, the American fort at Oswego on 6 May 1814, a number of critical military supplies destined for the US Navy frigate *Superior* remained out of British hands. Faced with how to best deliver these supplies to Sackett's, Master Commandant Melancthon Woolsey proposed to load the supplies onto bateaux. By sailing within reach of the American shoreline, rather than directly to Sackett's, the supplies stood a good chance to reaching their destination. To help guard the supplies, a company of US Rifles under Capt. Daniel Appling was assigned to the flotilla. A number of Oneida Iroquois, aligned with the American cause, also joined the group.

The Battle
As the British were patrolling the waterways from Oswego to Sackett's, the eighteen bateaux flotilla began its journey on the evening 28 May. The wind was poor and the squadron only reached the safety of Big Sandy Creek at 9:00 the following morning. One of the American boats, carrying a single 32-pdr. cannon and a 13-inch cable, had lagged behind and was captured by a British patrol. Excited by the possibility of further loot, the British pursued the Americans up Big Sandy Creek.

The Americans had driven their craft past a bend in the creek, hoping to hide the masts behind a small grove of trees. Seeing the American masts, the British disembarked men on both sides of their small seven boat flotilla. The British fired numerous times in the direction of the American craft with no response. The British continued to advance and when they got to within 30 yards, the Americans fired. The American fire was extremely accurate and the British gunner of the lead boat was hit seven times while the midshipman leading the British column was pierced with eleven balls. At the same time, the Native Americans attacked the British in the rear. Surrounded on all sides, the US Rifles then charged the British. The British surrendered.

It has been argued that the ambush was the best laid one throughout the war and one of the few recorded instances where rifles charged bayonets. The Americans suffered only two casualties while killing or capturing the entire British force, including the two

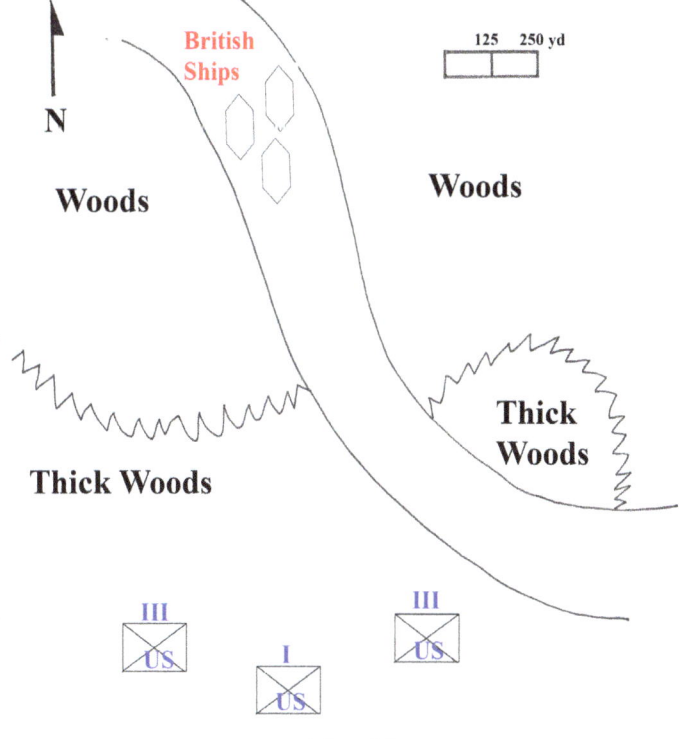

Map 21

gunboats and five barges. Additional American troops, including dragoons and a two-gun battery of 6-pdr. guns were dispatched to guard the shipment of supplies, but they did not participate in the engagement. The seventeen supply ships reached Sackett's Harbor without further incident.

Order of Battle
British - 188 men
Commanding Officers: Royal Navy Captains Popham and Spilsbury

Royal Marines, 2 units, 60 men in total
Sailors, 4 units, 120 men in total
2 gunboats
5 armed barges/bateaux

US - 210 men
Commanding Officer: Maj. Daniel Appling

First US Rifles, 84 men
Oneida Iroquois, 80 men
Armed seamen, 40 men
17 bateaux

45. US Rifles, author photo.

Game Notes

By all accounts, this was a perfectly executed ambush. The British captains disregarded explicit standing orders and followed the Americans up the creek. For the older gamers among us, the movie phrase "never get out of the boat" comes to mind. Though the flora and fauna that engulfed the mouth of Big Sandy Creek was clearly different from that which is found in Southeast Asia, the swampland that the British were trying to push through wasn't. As in Apocalypse Now, the technologically superior British spent most of their time chasing the forces of a third-rate world power and were getting beaten periodically in the process.

So how to make this ambush a successful wargame without including Huey's playing "Fortunate son"? Having played the scenario twice, a series of die rolls are needed to see if the Americans are able to spring their ambush successfully. It is possible that the British can spot the enemy, or an American unit would fire prematurely. The Americans got lucky in being able to coordinate their attack and the armed seamen, who at best were an untrained militia, did not do something stupid. In one of our games, the militia panicked early on, and the British were able to exploit the gap that formed in the American ambush.

If the British are able to survive the initial fire, then the British player needs to focus on either defending their position or breaking out of the swamp and try to reach open water. If the British do try and defend their position, then it is possible that the American reinforcements sent from Sackett's, specifically the two-gun battery or the cavalry, may enter the fray. If the British chose to withdraw all or part of their force, then some of their seamen need to crew the craft while the rest try to hold off the Americans.

Both British and American forces were raised on stories of the atrocities carried out by various Native Americans. At the beginning of the 19th century, scalping was still practiced and both European and American born recognized that they could be roasted alive over a fire. For this reason, the Royal Marines and seamen should always check their morale twice when being charged by Native Americans.

For their part, the Oneidas of 1814 were not the warriors of 17th or mid-18th century. Firearms were plentiful and their society had changed significantly. For all Native Americans, casualties were hard to replace and among the fragmented Iroquois, the loss of even a few men would be devastating. For this reason, the Oneida should stay in cover and fire at the British, rather than charge loaded muskets. No special morale check should be required of the Native Americans, however, when they charge an enemy fleeing from the battlefield.

For the American player, they are presented with the perfect plan, but plans often go out the window when the shooting begins. Units break and while the riflemen were known for their accuracy, the rifles of the time were slower to load. The Royal Marines can load and fire faster while their volleys can be devastating at close range. Charging an enemy is risky and the 1803 Harper Ferry rifles used by the riflemen did not mount a bayonet. The American charge could have easily failed, with the final result of the engagement unclear.

46. Oneidas, courtesy of Richard Watts.

Victory Conditions

The side which drives the enemy from the field wins the scenario. Consult your favorite rules to determine how many casualties a unit may suffer before becoming inactive on the battlefield.

Further Reading

Chester, Gregory, *Battle of Big Sandy, War of 1812.* (Watertown, NY. Published by the author. 2007)

Fredricksen, John D., *Green Coats and Glory: The United States Regiment of Riflemen, 1808-1821.* (Old Fort Niagara Association. Youngstown, New York. 2000)

Hough, Franklin, *A History of Jefferson County in the State of New York.* (Albany: Joel Munsell. 1854)

Slosek, Anthony, *Oswego and the War of 1812.* (Heritage Foundation of Oswego. Oswego, NY. 1989)

47. Royal Marine Firing Line, courtesy of Perry Miniatures.

Kingston, 21 October 1814
A What-If Scenario

Overview
Hypothetical naval engagement and American landing

Americans try to sink the HMS *St. Lawrence*

Introduction
By the start of 1814, the war of Lake Ontario had evolved from a haphazard affair to one fought by professionals. Though much has been written about how Gen. Scott trained the army and the militia, the same evolution occurred in the opposing navies. Gone were the days of hastily converted civilian transports and the Provincial Marine. Unfortunately for the Americans, the US Navy commander Chauncey remained timid as ever.

In early 1814, the British were able to complete two new frigates, *Prince Regent* and *Princess Charlotte*. In May, Yeo launched a raided across Lake Ontario and captured Oswego. Yeo then blockaded Sackett's Harbor for a few weeks. With the addition of two US Navy frigates, *Superior* and *Mohawk*, to the American flotilla, Chauncey then blockaded Yeo's squadron at Kingston. Launched on 10 September 1814, Chauncey's offer of battle was declined by the Royal Navy. On 19 October, lighting struck the *St. Lawrence*, killing seven crewmen and damaging a mast.

48. HMS *St Lawrence*, courtesy of John R. Lee.

The Battle
Recognizing the potential of the *St. Lawrence*, Chauncy approved of Midshipman James McGowan's plan to attack the First-Rate flagship. With a crew of eleven men, McGowan took a gig and one of Fulton's designed torpedoes. Though called at torpedo, the device was little more than a weighted barrel filled with gunpowder that would be attached to an enemy's hull. The resulting explosion would, in theory, sink the warship.

McGowan's gig ran from Sackett's Harbor up the US coast, to Wolfe Island and into the Kingston channel. There McGowan's gig ran into a British patrol boat, which they captured. The captured seamen convinced McGowan that the *St. Lawrence* was no longer in Kingston. Frustrated by his lost opportunity, McGowan returned to Sackett's.

With the British clearly having gained an advantage, Chauncey did not leave the safety of Sackett's Harbor, where he waited for the delivery of two American ships of the line and a new frigate. These never were launched, as the war came to an end. The warships of both nations were then left to rot or sold for scrape.

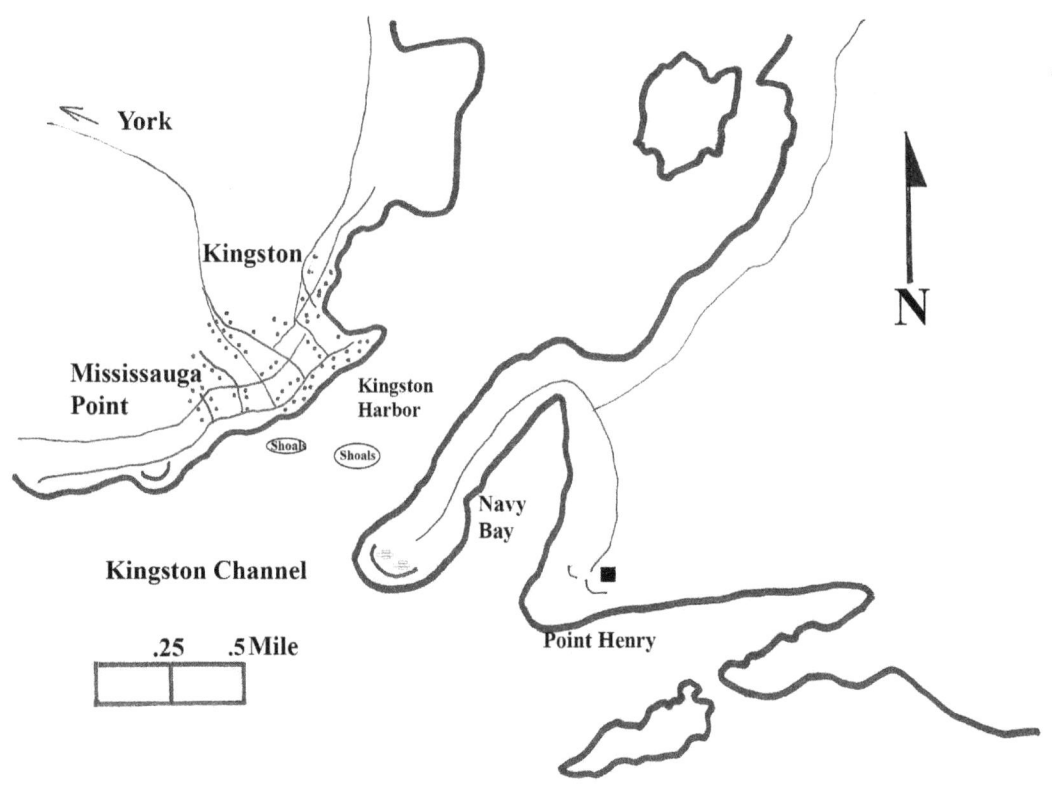

Map 22

Order of Battle

British

Commanding Officer: Commodore Sir James Lucas Yeo

HMS *St. Lawrence,* ship of the line, twenty-eight 32-pdrs., forty 24-pdrs., four 68-pdr. carronades, thirty-two 32-pdr. carronades
HMS *Prince Regent,* frigate, thirty-two 24-pdrs., four 68-pdr. carronades, twenty-two 32-pdr. carronades
HMS *Princess Charlotte,* frigate, twenty-six 24-pdrs., two 68-pdr. carronades, fourteen 32-pdr. carronades
HMS *Montreal,* sloop, seven 24-pdrs., eighteen 18-pdrs.
HMS *Niagara,* sloop, two 12-pdrs., twenty 32-pdr. carronades
HMS *Charwell,* brig, two 12-pdrs., fourteen 32-pdr. carronades
HMS *Star,* brig, two 12-pdrs., fourteen 32-pdr. carronades
HMS *Netley,* brig, two 12-pdrs., fourteen 24-pdr. carronades
HMS *Magnet,* brig, two 12-pdrs., twelve 24-pdr. carronades

US

USS *Superior,* frigate, thirty 32-pdrs., two 24-pdrs., twenty-six 42-pdr. carronades
USS *Mohawk,* frigate, twenty-six 24-pdrs., two 18-pdrs., fourteen 32-pdr. carronades
USS *General Pike*, frigate, twenty-six 24-pdrs., two 24-pdr. chase guns
USS *Madison*, corvette, two 12-pdrs., twenty-two 32pdr. carronades
USS *Jones*, brig, two 12-pdrs., twenty 42-pdr. carronades
USS *Jefferson*, brig, two 12-pdrs., twenty 42-pdr. carronades

USS *Sylph*, brig, two 12-pdrs., fourteen 24-pdr. carronades
USS *Oneida*, brig, two 12-pdrs., fourteen 24-pdr. carronades

49. 1st Rate and Frigates on Lake Ontario, October 1814, courtesy of Chris Ngiau.

Game Notes

This scenario definitely falls into the scope of what ifs – What if the British came out of Kingston harbor and took up Chauncey's challenge? What if Chauncey devoted more resources to the torpedo attack? While Chauncey was replaced as the US Naval Commander before the war came to an end, what if an aggressive commander like Perry or Macdonough, had taken command earlier?

As wargamers, we are attracted to the biggest and the best. WW2 gamers love the King Tiger and Napoleonic gamers field a lot of 12-pdr. cannons. The *St. Lawrence* clearly falls into this bigger is cooler category. Yet untried in battle, this ship-of-the-line carried a lot of firepower, but Yeo needed to test the *St. Lawrence's* sailing abilities before risking her in battle.

In this hypothetical scenario, Chauncey has been replaced by the newly appointed Perry. Knowing this may be his last opportunity to take the war to Canada, Perry's goal is nothing short of sinking the *St Lawrence*, even if it means attacking Kingston. Perry's plan is bold – take everything that floats and can mount a gun and throw it at Kingston. In the confusion, a series of small craft will make a run at the *St. Lawrence*. If the small boats are unable to damage or sink the *St. Lawrence* outright, maybe the two US Navy frigates can.

For this scenario, the Royal Navy has returned to port to take on supplies and men. In the early morning fog, the British see something moving. Can it be that the American flotilla has come out to play? Jolly good! Let's give these Johnnie's a damn good thrashing!

Fulton's torpedoes are in reality mines. With the British watching the two American frigates, the three tiny American gigs are rowing toward the *St. Lawrence*. The American gigs must row within two inches of the *St. Lawrence* and then throw a harpoon into the ship. The *St. Lawrence,* like all other warships sailing on the Great Lakes, was not copper sheeted and the hand thrown harpoon should be able to attach the torpedo to the target. We have played this scenario with the British being able to sink the gigs before they have a chance to attach the torpedoes to the side of the warship and in each case the British player made a point of concentrating their fire on the small craft. In more recent tabletop recreations, we added a house rule that does not allow the Royal Navy to depress their cannon low enough to target the American gigs. This house rule not only adds to the excitement/tension, but historically small craft would only be targeted after the larger threats have been neutralized.

It will take each gig four turns to row into position and a complete turn to try and harpoon the side of the vessel. Harpooning is not automatic, and there is a 50% chance of success on each attempt – a roll of a 4, 5, or 6 on a D6 is needed. If a gig fails to harpoon the targeted enemy vessel, they can try again next turn.

Once the target ship is harpooned, it is unclear if the torpedo actually detonates. Detonation would occur at the start of the following turn. In the early 1800s, Fulton was able to sink a few test craft, including the 200-ton brig *Dorothea*, but that was done in Europe. Again, a die roll on a D6 determines the result, with a 5 or a 6 being needed for a detonation. If a detonation occurs, a critical damage roll needs to be made. Most naval rule sets have a critical hit table and if one is using Warlord's Black Seas ruleset, use the Critical Table – Hull, which is found on page 20. Once a torpedo is detonated, regardless of the damage caused, the Royal Navy can target the American gigs.

In the 1780s, the British began to develop Kingston and they constructed new shipyards and naval installations on the eastern shore of the Cataraqui River. After

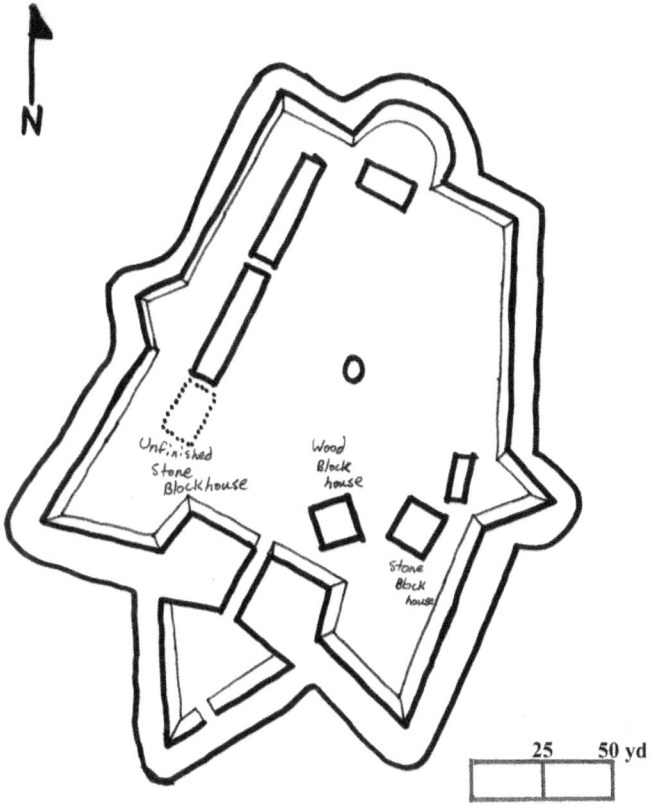

50. Plan of Ft Henry, drawn by Christian Wright.

almost three years of war, the defenses at Kingston were formidable. Men, fortifications, and cannon all protected the Royal Navy's main supply base on Lake Ontario. The town of Kingston itself was protected by a log palisade wall and six blockhouses, while batteries and additional blockhouses protected Point Frederick and Point Henry. The naval dockyard constructed many of the British warships sailing on Lake Ontario and in the late fall of 1814, two new ships of the line, the HMS *Wolfe* and the HMS *Canada* were under construction. Chauncey's halfhearted assault on Kingston in late 1812 resulted in the British strengthening their defenses and making sure that the area was garrisoned by both British regulars and militia.

Chauncey was always dismissive of launching a raid on Kingston and would over-estimate the area's defenses. Though a large number of American troops would be available to attack Kingston – Izard had 4,500 trained men from the Right Division available at Sackett's Harbor on 16 September – Chauncey sent them to Ft. Erie. Though Chauncey ignored Izard's request to attack Kingston, what if he did not? Rather than sending Izard to the Niagara Peninsula, what if a new naval commander decided to take this force and attack Kingston?

In this final variant, Izard's 4,500 men go along to Kingston. Izard was a fine administrator and was rightly frustrated with the situation, but he was not an overtly aggressive commander like Brown or Scott. Though Izard's eventual abandonment of Ft. Erie to the enemy was a controversial decision at the time, most scholars agree that there was little point of holding on to the real-estate without adequate naval support. If Izard leads an attack on Kingston, he would most likely stop his assault if he ran into serious opposition. For game purposes, if the American army suffers more than twenty percent casualties in an attack, Izard will order the entire army to withdraw.

If one were to continue to play the what-if scenario, one would only be able to land Izard's division after the American Navy had gain tactical superiority over the Royal Navy. The purpose of the raid is to destroy British ship building facilities and any boats under construction. Rules for landing and destroying public property were discussed previously in the sections on the American attacks on Ft. George and on York, so rather than repeat them

51. Fulton's Torpedo Boat in action, courtesy of John R. Lee

again, the reader is asked to look at these scenarios which appear earlier in this work.

For this hypothetical raid, the following forces are suggested. Though Izard had an attached battery of guns and Eustis's Second Light Dragoon Regiment as part of this command, historically these were not transported by the Navy to the Genesee River and thus are not included in this hypothetical attack. For artillery support, the US Navy can provide fire support during the landing, either from the six gunboats assigned to the attack (each mounting a single long 24-pdr. gun in the front) or from any boats in the American flotilla still afloat after the attack on the *St. Lawrence*.

As the focus of the Americans is the British base on Navy Bay, the Americans need to land and capture Ft. Frederick. Once that has been neutralized, the Americans then can move on to destroy the naval docks and yards. The game plays until the Americans are driven off or they reach Navy Bay before suffering 20% casualties.

Order of Battle

Map 23

British - 1,868 men (not including any militia), 591 men stationed at Points Henry and Frederick
Commanding Officer: Maj. Gen. Louis de Watteville

9th Regiment of Foot, *400 men*
37th Regiment of Foot, *360 men*
104th Regiment of Foot, *300 men*
De Watteville's Regiment, *600 men*
Canadian Fencibles, *168 men*
Glengarry Light Infantry Fencibles, *40 men*

Ft. Henry (at Point Henry)
Earthen wall fort with two bastions and a ravelin, protected by a parapet. In the center there was a wooden blockhouse and a recently completed stone blockhouse (No. 1 Tower). A second stone tower was still under construction.

17 guns, including:
Five 24-pdrs.
One 18-pdr. facing Cedar Creek, mounted on a traversing platform
One 18-pdr. carronade facing Point Frederick, mounted on a traversing platform
One 8-in. brass howitzer
Four 9-pdrs. (one in the wooden blockhouse)
One 6-pdr. (in the wooden blockhouse)
One 4-pdr.
One 24-pdr. carronade (on the parapet of the ravelin)
One 18-pdr. carronade

Point Frederick
Two batteries
Wooden blockhouse

Mississauga Point
Battery

Kingston
Log palisade wall and blockhouses

US 4,500 men
Commanding Officer: Maj. Gen. George Izard

1st Brigade
Brig. Daniel Bissell
Fifth US Infantry Regiment, *500 men*
Fourteenth US Infantry Regiment, *500 men*
Fifteenth US Infantry Regiment, *500 men*
Sixteenth US Infantry Regiment, *500 men*

2nd Brigade
Brig. Thomas Adams Smith
Fourth US Infantry Regiment, *500 men*
Tenth US Infantry Regiment, *500 men*
Twelfth US Infantry Regiment, *500 men*
Seventeenth US Infantry Regiment, *500 men*

Naval Landing Party
US Marines, 300 men
US Seamen, 200 men

52. US Infantry, courtesy of Knuckleduster Miniatures.

Victory Conditions
    Americans need to drive the Royal Navy from Lake Ontario. If they Americans accomplish this, they will have a much stronger position at the ongoing peace negotiations. Without control of Lake Ontario, the British may even withdraw from Upper Canada. If the Americans fail, the timeline is restored, and nothing changes. The game plays until the American force withdraws or is destroyed.

Further Reading

Cary, Henry, Early Works: Preliminary Glimpses of the First Military Complex at Point Henry, Ontario, 1812-1817. (*Ontario Archaeology*. Vol. 76., No. 4. 2003)

Lardas, Mark, *Great Lakes Warships 1812-1815*. (Osprey Publishing. Oxford. 2012)

Malcomson, Robert, HMS St Lawrence: The Freshwater First-Rate. *The Mariner's Mirror*. Vol. 83, No. 4. (November). 1997

Mecredy, Stephen D., Crisis Confounding Construction: the History of Point Henry During the War of 1812. (*Historic Kingston*. Vol. 33., 1985)

53. Royal Artillery, Courtesy of Perry Miniatures.

Niagara Peninsula/ Western Lake Ontario

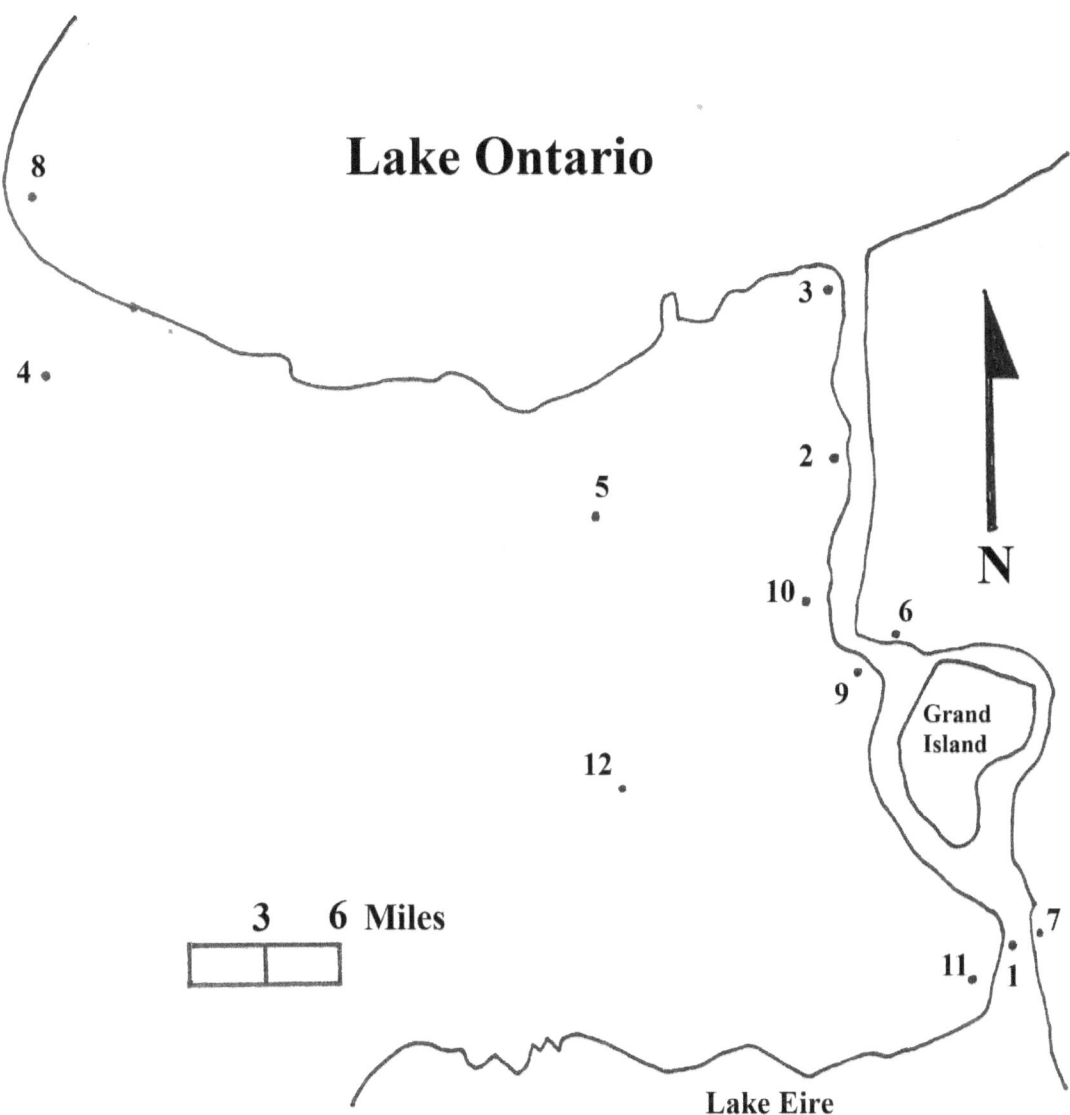

1. HMS Caledonia & Detroit 2. Queenston Heights 3. Ft. George 4. Stoney Creek
5. Beaver Dams 6. Ft. Schlosser 7. Black Rock 8. Burlington Races 9. Chippawa
10. Lundy's Lane 11. Ft. Erie 12. Cooks Mill

Map 24

HMS *Caledonia* and HMS *Detroit*, 1 October 1812

54. Capt. Elliot by David Edwin 1813, Naval History and Heritage Command

Overview
Small Skirmish - less than 500 participants

US troops cross the Niagara River and capture two brigs. One is sailed safely back to Black Rock while the other ran aground and was set on fire before she could be recaptured.

Introduction

The early months of the War of 1812 were not going well for the Americans. The elaborate three-pronged attack into Canada failed on all fronts. The eastern attack on Montreal never really materialized and the one assault across the Niagara on Queenston was an American disaster. Far worse for the Americans was the western theater, where the British and their Native American allies not only stopped the half-hearted American advance into Upper Canada, but the forces loyal to the crown had captured much of Michigan. While one may choose to highlight the incompetence of the American military establishment and the delusional schemes of the US War Department, accounts of individual bravery and competence were also present during the first year of the war. On the British side, the feats of Gen. Brock have long been recognized but those American junior officers are rarely acknowledged. At the same Battle of Queenston Heights, for example, the efforts of junior officers such as John Wool and Winfield Scott were both proper and professional. The actions of Lt. Jesse D. Elliott, USN, are even less recognized, even among those who have an interest in the War of 1812.

In the late summer of 1812, Lt. Jesse D. Elliott was sent by Commodore Chauncey to build a fleet on Lake Erie. The young Lt. was to consult with Gen. Van Rensselaer and purchase merchant vessels to be converted to vessels of war. In addition, Elliott was to build new warships and a naval base with quarters for 300 men. Elliott chose the settlement at Black Rock on the Niagara River as his home port. As Elliot was busy in putting together a US Navy presence on Lake Erie, he learned that two craft, the *HMS Caledonia* and the *HMS Detroit* had arrived at Ft. Erie, which was just across the Niagara River on the Canadian side.

The *HMS Caledonia* was built in 1807 for the Canadian North West Company and at the start of the war was pressed into British service with the Provincial Marine. The *Caledonia* took part in the capture of Ft. Mackinac and Detroit. Mounting two long 24-pdrs., one 32-pdr. carronade, and four smaller 4-pdr. swivel guns, the boat was one of the best armed craft west of the Niagara Falls.

The *HMS Detroit* was the former American built *USS Adams*. The *Adams* was built at River Rouge (Michigan) launched in 1798. Before the start of the war, the *Adams* was the only American military vessel in operation on the western Great Lakes. When Gen. Hull surrendered Detroit to the British, the *Adams* was taken by the Provincial Marine and put into service with the British. Renamed the *HMS Detroit,* the six-gun brig, along with the *Caledonia,* carried supplies to the Niagara peninsula in anticipation of an American attack.

With no warships with which to challenge the British naval superiority on Lake Erie, Elliott put together a plan to shift the balance of power to the US side. Taking command of the still unarmed recently arrived Navy Seaman and borrowing fifty men of the Second Artillery Regiment, Elliott scrounged enough pistols, swords, and sabers to arm his group. Joined by a few local volunteers, Elliott's party rowed out in two boats to do battle with

The Raid

The British were lax in their guard duties and Elliot's men were able to board the two craft without alerting their foes. The *Detroit* was captured without much opposition and the *Caledonia* fell into American hands following a short but brief melee. As the wind was too light to sail the prizes into Lake Erie, the boats were run downstream in the Niagara River.

With the alarm given, the British forces began to respond. The British batteries at Ft. Erie began to fire on both craft and the *Detroit* became embroiled in an artillery duel. The *Caledonia* reached the safety of an American gun battery between Buffalo and Black Rock, but the *Detroit* became grounded on a shoal. Attempts to haul the *Detroit* to the American shore were unsuccessful and Elliot was forced to cut her adrift. Without a pilot, the *Detroit* continued downriver, eventually running aground on west side of Squaw Island.

In breaking daylight, the British launched a party of men from the 49th Foot in an effort to recover the *Detroit*. From the American side, a party of local militia under Ebenezer Walden crossed over to Squaw Island in a scow to lay claim to the *Detroit*. The Buffalo Militia brought with them a six-pdr. field piece, which helped them temporarily drive off the 49th Foot. At this point Gen. Sir Isaac Brock arrived on the scene and began to make plans to recapture the *Detroit,* but before they could be put into play, a party of men from the Fifth US Infantry boarded the *Detroit* and set her ablaze.

The *Caledonia* was a rich prize and the furs she carried were valued at over 200,000 dollars. As a result of the action, the American prisoners on both boats were set free and the *Caledonia* for a time formed the nucleus of the American squadron on Lake Erie.

Order of Battle

British
At Ft. Erie
HMS Caledonia
Mr. Irvine
12 men
10 US prisoners

HMS Detroit
Lt. Rolette
56 men
30 US prisoners

Land Battery
Maj. Ormsby

Reinforcements
Gen. Sir Isaac Brock
49th Foot, *450 men*
HMS Lady Prevost

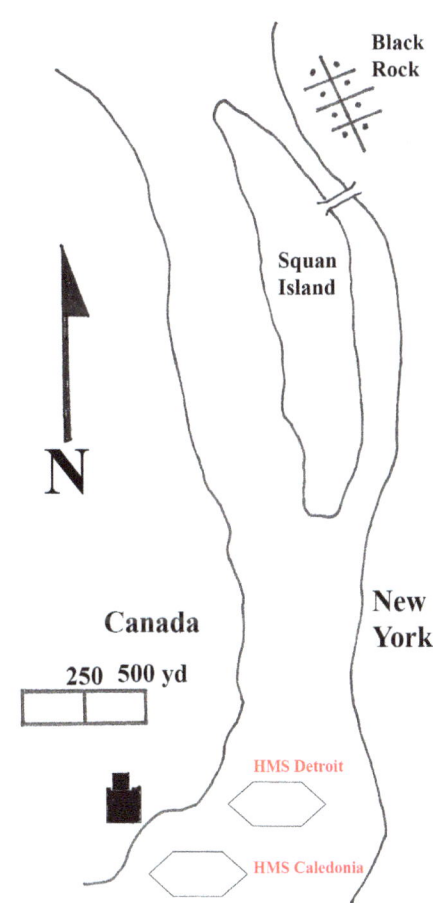
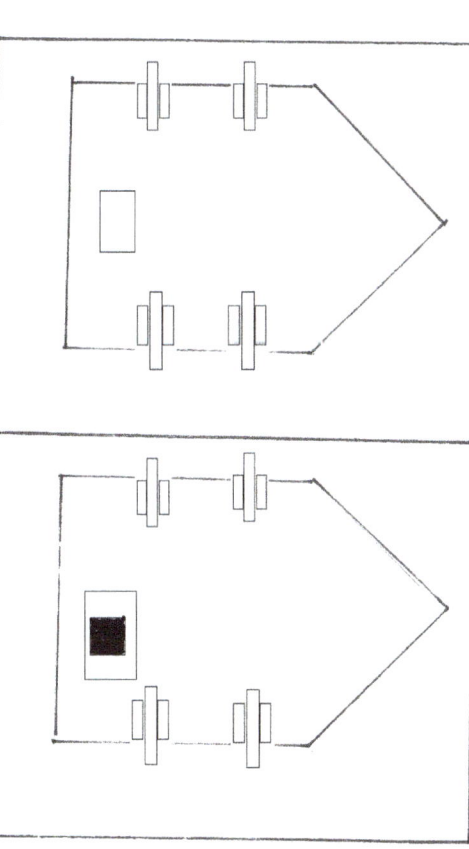

Map 25

US
Commanding Officer: Lt. Jesse D. Elliott, USN

Initial Raiding party, 124 men
Lt. Jesse D. Elliott, USN
seamen, 50 men
Capt. Towson
Second Artillery Regiment, 50 men
Two large boats

Reinforcements
New York State Militia, *100 men*
6-pdr. cannon, *12 men*
Fifth US Infantry, *200 men*

Game Notes

The American action to capture the two boats almost sounds like an adventure movie with a daring night attack followed by a desperate run back to friendly lines. With such a vivid plot line, the action makes for a great skirmish game.

We have often recreated the attack on boat by a boarding party and it makes for a fun action. Each boat is considered to have two levels, a top deck and three sections below deck. The three sections below deck can only be accessed from the top deck. Before the Americans launch their attack, the British player notes where he will place his crew, with only a 1/3 of the defenders being on the top deck. Before the attacker can attempt to grapple with the defending craft, the British player needs to roll to see if the lookouts spot the launches. On a roll of a 1 or a 2 on a D6, the British see the Americans and all hands are called out to repel boarders. If the Americans are not spotted, they are able to land and all defenders are considered to be disordered.

Once the Americans secure the top deck, they must quickly capture the rest of the ship. The American player then identifies which units enter which of the three holds. The following turn the Americans and British melee in each section. If the British win even one of the three actions, they are free to attack the Americans or can try to sabotage the ship to keep it from falling into enemy hands.

If the Americans are successful in capturing a boat, then they need to see if they can escape. The Americans must dice for wind, needing a 1, 2, or 3 on a D6. If they do not have the wind, the Americans must try to send the boats down river. This is achieved by raising anchors and moving 10 inches every turn. The current is tricky and on a 1 or a 2 on a D6 die roll, the boat will drift 6 inches to the left or the right of the ship's bow. It takes six turns to drift completely to the American side of the river.

As the captured boats drift past the batteries of Ft. Erie, the British player may fire the shore battery at the targets. The British may also send out their reserves to recapture the boats if they become stranded on one of the islands in the Niagara River. If a boat becomes stranded on an island, the New York State Militia may also move out to help.

The game plays for ten turns.

55. Small boats, author's photo.

Victory Conditions

Major US victory – capture both boats and bring them to the US side of the Niagara River.
Minor US victory – bring one boat to the US side of the Niagara River.
Draw – The boats are destroyed.
British victory - Recapture both boats

56. Excelsior Advances, 40mm New York State Militia, author's photo.

Further Reading

Lossing, Benson J., *Pictorial Field-Book of the War of 1812*. (Harper& Brothers Publishing. New York. 1869)

Collins, Gilbert, Guidebook to the Historical Sites of the War of 1812. (Dundurn Press. Toronto. 2004)

Roosevelt, Theodore, The Naval War of 1812. 3rd edtion (G.P. Putnam's Sons, New York, 1883)

57. British Infantry, courtesy of Sash and Saber Castings.

Queenston Heights, 13 October 1812

Overview
Battle - 3,000 participants
Small Skirmish - 60 men

The US tries to invade Upper Canada but is defeated. The engagement was one of the first major battles in the war.

Introduction

At the start of the war, the Americans had drafted elaborate plans to bring the British colony of Upper and Lower Canada into the American fold. Even after Hull's surrender at Detroit early that year, many Americans continued to believe they could easily capture Canada. As the Great Lakes laid between the two warring parties, invasion routes were limited to the thin strips of land between the Lakes.

A plan formulated by Washington called for a four-pronged attack across the entire frontier. Maj. Gen. Stephen Van Rensselaer was to launch an attack across the Niagara River which separated Lake Erie from Lake Ontario. Van Rensselaer's advance was to sever the British forces operating to the west from their eastern supply lines and equally important, would draw the British to the Niagara Peninsula. Such an advance would then allow Dearborn to move down Lake Champlain and up the St. Lawrence to Montreal and Quebec.

Van Rensselaer was appointed by the State of New York and was immediately at odds with lower ranking regular US army officers. Assisted by cousin, the adjutant general Lt. Col. Solomon Van Rensselaer, the two labored to put together a fighting force. The New York State Militia grew tired of waiting and asked to be given an order to attack or they would go home. The arrival of Brig. Gen. Alexander Smyth, now in charge of a brigade of 1,700 regulars, did little to help the situation. Smyth chose to ignore any order from Van Rensselaer and in the end did not participate in the upcoming attack.

The lack of boats hampered Van Rensselaer's modified plan, but the general recognized if he did not attack, his political enemies would exploit his delay to their advantage. On 10 October boats arrived at Lewiston and the American forces began preparing for their immanent assault. British observers on the other side of the river reported the American activities to their commander, Gen. Isaac Brock. The loss of oars and oarlocks, along with a heavy rainstorm, delayed the start of the US assault until the early hours of the 13th.

Gen. Brock had expected an American assault to focus on either Ft. Erie or Ft. George but being aware of the American build up at Lewistown, he had actively strengthened the defenses all along the Niagara River. Facing the American town of Lewiston, Brock ordered the construction of a heavy battery overlooking the river. The Americans also brought two 18-pdr. guns and a heavy mortar to support their landing. Though he had prepared as much as possible, an American attack at Queenston could easily be a diversion, for a large number of American troops remained at the source and the mouth of the Niagara River.

The Battle

Solomon Van Rensselaer, the man leading the assault, split his force into waves, with him personally leading the first wave. At 0400 hours, the first wave of American militia and regulars departed New York for Canada. Of the thirteen boats that tried to get across the river, three failed to do so. Upon landing on a gravel beach, the initial landing force was attacked by two companies of British infantry. Though the attack was beaten back, Solomon Van Rensselaer was wounded. Command then fell to the wounded Capt. John Wool.

The British artillery continued to pummel the American embarkation point as boats filled with wounded returned to the American shore. Fresh troops then filled the boats and then tried to cross the river again. Lt. Col. John Fenwick attempted to land his four boatloads of troops further downstream, but this force was decimated by a combination of British grapeshot and musket fire.

Back at the original landing site, the American troops had secured the beachhead, but with the loss of their commander, no clear battle plan existed. Learning of a rarely used trail that led up the steep 300-foot escarpment, Wool and his men reached its summit. From this position, the US troops were able to assault the rear of the 18-pdr. gun position and secure it.

Learning of the American attack, the British Commander Gen. Sir Isaac Brock arrived on the battlefield to assess the situation. Recognizing the importance of this gun in limiting the flow of US troops appearing on the Canadian shore, Brock led an assault up the hill on the Americans. A well-placed musket shot cut Brock down and the British assault faltered. A second British attack led by Lt. Col. John Macdonnell failed, with every officer of the Crown being killed.

With the silencing of the British 18-pdr. gun, there was a lull in the fighting. Americans continued to haphazardly trickle in additional troops. Militia Brig. Gen. William Wadsworth crossed into Canada, but many of the New York State Militia did not cross in the blood covered boats. Lt. Col. Winfield Scott could not get his 6-pdr. guns loaded into the craft but was allowed to cross. Arriving on the heights, Scott assumed command and the wounded Wool was evacuated to the American side of the river.

At this time, Maj. Gen. Roger Sheaffe brought up fresh British troops from Ft. George. To screen his advance, Native Americans advanced toward the American lines. The Mohawk's piercing yells unnerved the raw American troops still waiting to embark. Scott then organized and led a bayonet charge, which drove them from the field. The American troops on the escarpment were hungry and mostly out of ammunition.

On the march, Sheaffe's troops from Ft. George were joined by more men from Chippawa. At about 1500, the British force, which now outnumbered the Americans, began to advance. Scott then directed a fighting withdrawal to the embarkation point. Native Americans pursued and scalped the retreating US troops. At the landing spot, no boats were waiting. Scott attempted to hold off Native American attacks. Sometime later, Wadsworth surrendered his sword to Sheaffe and the battle was over.

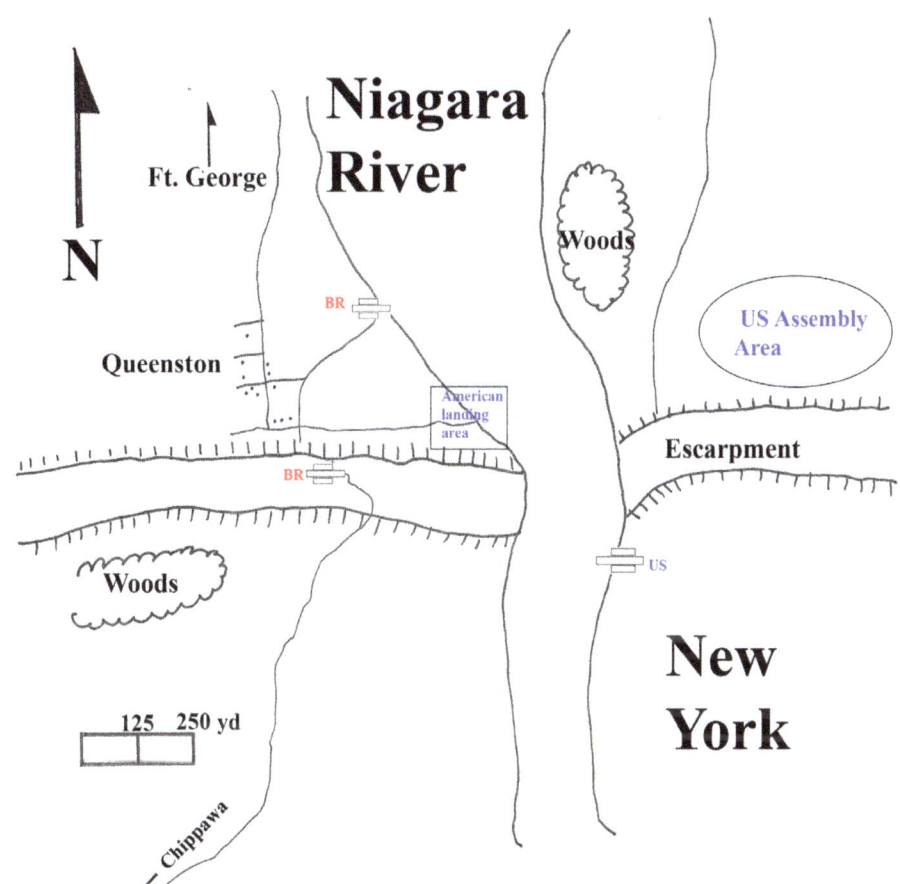

Map 26

Order of Battle

British - 1,500 men
Commanding Officer: Maj. Gen. Sir Isaac Brock

At Queenston at the time of the Landing, 456 men
49th Regiment of Foot, two companies, 200 men
Royal Artillery, 6 men
5th Lincoln Regiment (militia), 100 men
2nd York Regiment (militia), 80 men
3rd York Regiment (militia), 30 men
Lincoln Artillery (militia), 40 men

In the Area
Joe Brant and Cpt. John Norton
Native American Mohawk, 80 men

Reinforcements from Fort George and Fort Erie, 913 men
Maj. Gen. Roger Hale Sheaffe
41st Regiment of Foot, 380 men
Royal Artillery, 33 men
Militia, 500 men

US 3,500 in the area, with about 1350 taking part in the battle
Commanding Officer: Maj. Gen. Stephen Van Rensselaer

1st Wave
Lt. Col. Solomon Van Rensselaer, Capt. John Wool
Thirteenth US Infantry Regiment, 260 men
Selected New York State Militia, 200 men
Volunteer New York State Riflemen, 24 men

2nd Wave
Lt. Col. John Fenwick
US artillerymen fighting as infantry, 150 men
New York State militia, 200 men
Volunteer New York State Riflemen, 46 men

58. Isaac Brock Action Figure, author's photo.

3rd Wave
Maj. James Mullany
Twenty-Third US Infantry, 200 men
New York State Militia, 160 men
Volunteer New York State Riflemen, *28 men*

4th Wave
Lt. Col. Winfield Scott
New York State Militia, *300 men*
Volunteer New York State Riflemen, *22 men*

5th Wave
Brig. Gen. William Wadsworth
New York State Militia, *100 men*

6th Wave
New York State Militia, *200 men*

Game Notes

Gamers have periodically recreated the Battle of Queenston Heights with various degrees of success. For some, Brock is looked to as a brilliant hero cut down in his prime while Sheaffe, the true victor of the battle, rarely receives the credit for stopping the invasion. Both Stephen and Solomon Van Rensselaer carried themselves admirably but were overwhelmed and unprepared for the logistical nightmare of leading a contested river crossing. Some units, despite their raw nature, demonstrated competency above their level of training, as did the leadership skills of a number of officers on both sides.

Given the near mythological reverence for this battle in Canadian history, both the entire battle and segments of it have been recreated on the tabletop. When trying to recreate the entire US attack, three issues are critical. The first - how, where and when will troops appear on the battlefield. In our games, we allow the 1st wave to arrive at the designated landing site, but for each additional wave, 2 D6 are rolled to see how much further downstream from the original landing site each subsequent wave lands. Each wave should land each turn thereafter, but this did not go according to plan. A D6 is rolled to duplicate the issues of the boats returning to the embarkation point, loading troops and supplies, and making it back across the river to the Canadian side. On a roll of a 4, 5, or 6, the landing of this wave (and subsequently all additional waves) is delayed for a turn.

59. View from atop the Niagara Escarpment toward the American Embarkation Area, author's photo.

The second issue is how each unit will behave when they first come under fire. The actions of the two British regular line units, the 41st and 49th do not need to be tested before they take part in a battle, but for all others it remains unclear how they will act. Individually, riflemen were chosen for their proficiency with firearms, but that does not necessarily translate into them acting as a disciplined unit.

To replicate a unit's behavior the first time it faces the enemy, a D6 die role determines its fate. On a roll of a 1 or 2, the unit behaves normally. On a roll of a 3 or a 4, the unit may fire and reload, but cannot advance. On a roll of a 5, the unit falls back facing the enemy. On a D6, the unit will attempt to charge the nearest visible enemy unit. If the charging unit fails to make contact with its target, the unit is considered to be disordered.

The third issue is effect of Native Americans on the battlefield. The recurring nightmare of Native American raids played havoc with the militia, they may also panic at the sounds of Mohawk warriors. American militia waiting to cross into Canada observed the boatloads of wounded working their way through their camp on their way toward treatment unnerved many, but the war cries heard from across the river resulted in many a militiaman losing their courage. Though the constitutionality of New York State Militia serving in Canada may have been a philosophical debate earlier, the stark realities of combat pushed this thinly veiled pretense out into the forefront.

The Native Americans should appear half-way through the game. At this point, any militia troops on the American side of the river need to dice to see if they decide to stay in New York or embark to Canada. On a roll of a 4, 5, or 6 on a D6, no further American troops will make it into Canada. If Scott has not made it to Canada and the US troops have stopped reinforcing the landing, a roll of a 5 or a 6 on an additional D6 die roll will bring him directly to the Canadian side of the river at the designated landing zone.

The game begins with the 1st wave landing and coming into contact with the British troops at Queenston. The 18-pdr. cannon on top of the escarpment cannot be depressed low enough to fire at the landing site. Gen, Brock should appear at the start of turn 4, the Native Americans at turn 5 and the Sheaffe at turn 6. It should take

60. US Infantry in Early Uniforms, authors photo.

the Americans 1 turn to climb the unprotected path to the top of the escarpment. The game plays for 10 turns.

If, however, one wanted to game a particular aspect of the battle, such as Brock's futile attempt to recapture the 18-pdr. cannon or Scott's escape from the redoubt to the landing site, one can do so with a much smaller number of figures and far different goals.

Brock's death cemented him into the pantheon of Canadian national heroes, a role he continues to hold. Brock has his own action figure, an accomplishment no American commander comes close to replicating. Gaming Brock's assault to recapture the gun, however, can be often futile. Though Brock's force is understrength when attacking the Thirteenth US Infantry, the issues with this tabletop recreation is the game can become one of "Kill Brock". Historically, the number of officer casualties on both sides was extremely high and with Brock's passing, his attack failed. The follow up attack resulted in all British officers becoming casualties. While one could add a rule which does not allow troops to target specific officers, such a rules amendment would be contrary to the historical reality. In three different games played by the author, Brock was killed - either by himself or when attached to a unit.

Victory Conditions

If one was to play out the entire invasion, the British need to send the Americans back across the Niagara River. If the Americans, however, are able to establish and hold a beachhead, they achieve a minor victory. If the Americans are able to push inland and get a unit in good order off the British edge of the table, they achieve a major victory.

Small Skirmish

A more even and less political section of the game would be recreating the American fighting retreat to the landing site. American units are at best company size and many individuals are fleeing for their lives. Some troops who attempted to move down the escarpment, lost their footing and plummet to their death, while others were killed and scalped by the Mohawk who had pierced the American line.

One recreation of this event, played at a one-to-one scale, included Lt. Col. Scott and nineteen Americans being pursued by an equal number of Native Americans. To keep the Americans moving to the embarkation site, a second group of British troops followed in the Mohawk footsteps starting at the end of turn 3. Native Americans should be able to move faster than the Americans. If your rules do not include a movement bonus for native troops, give them an additional quarter movement advantage. For example, if an American figure has a movement of 8 inches, a Native American warrior could move 10 inches.

Having climbed the escarpment many times in my youth, it is rather easy to lose one's footing and slip. Any figure that tries to run should require a test to see if they slip and if so, what damage they suffer. A D6 is rolled to see if a figure falls – on a roll of a 1, 2, or 3 the figure does not slip but if a 4, 5, or 6 is rolled, then an additional D6 is required to see what happens. On a roll of a 1 or 2, there is no effect, on a 3 or a 4 they fall down and must use their next turn to get up, on a 5 they can no longer move by themselves and on a 6 they plummet to their death while falling off the escarpment like in an Indiana Jones movie.

Though most Native Americans in this region have been using firearms since the middle of the 17th century, interviews conducted by white men in the later years of the 19th century underlined the continued use of traditional melee weapons. For game purposes, our Native American warriors are armed with both.

To reflect the lack of ammunition among the American troops, and to keep things interesting, the number of cartridges available to each American was established to be between 1 and 6, with a D6 roll determining what was left after the first shot. Only after firing their first shot did the American player roll a D6 and learn how many more shots remained in their cartridge box. American figures can share cartridges if they are within an inch of each other. All US infantry are armed with muskets and bayonets. Lt. Col. Scott is initially armed with a pistol and sword but can pick up a musket from a killed or wounded US soldier. The game ends when the American player reaches the landing site.

Victory Conditions

If one were playing the American retreat, then a minor victory is achieved when the US bring more than half of their force back to the landing site and gain a major victory when they lose no more than quarter of their force. Conversely, the British achieve a major victory when they destroy more than half of the invaders and gain a minor victory when they vanquish three fourths of the Americans. The game plays until the last American reaches the landing zone or is killed.

Further Reading

Barbuto, Richard V., *Staff Ride Handbook for the 1812 Campaigns, 1812-1814.* (Old Fort Niagara Association. Youngstown, NY. 2016)

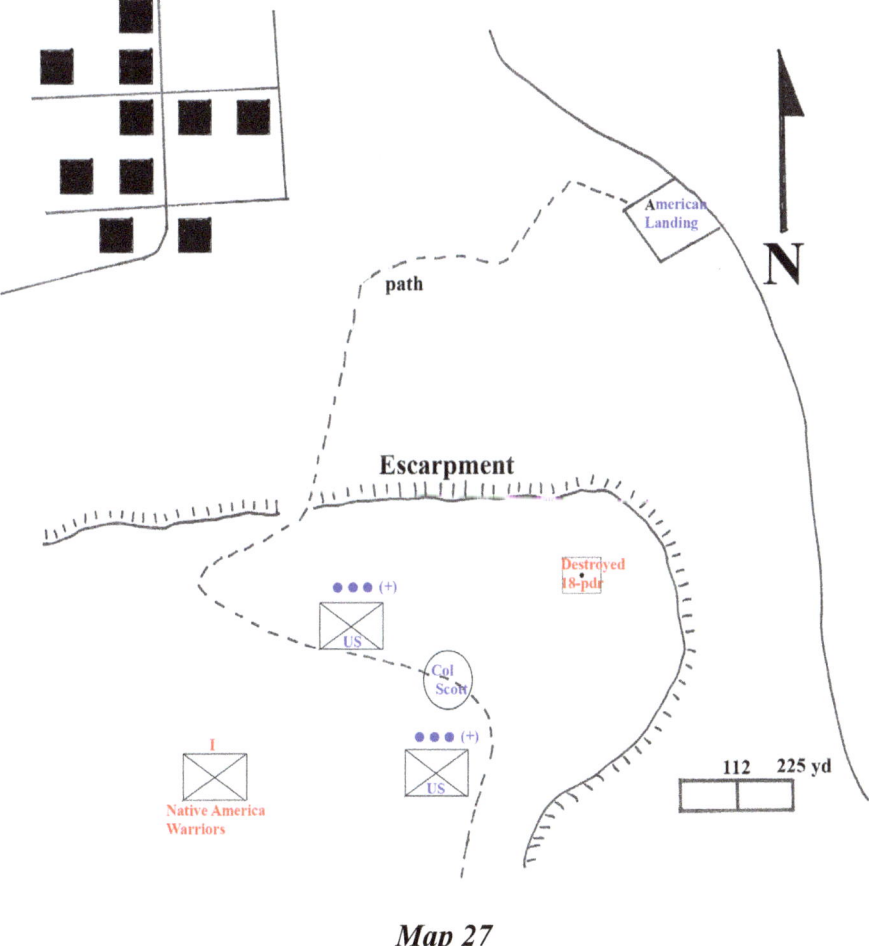

Map 27

Cruikshank, E. A., *The Battle of Queenston Heights*. (Lundy's Lane Historical Society. Niagara Falls. 1891)

Malcomson, Robert, *A Very Brilliant Affair: The Battle of Queenston Heights, 1812*. (Robin Brass Studio. Toronto. 2003)

Ft. George, 27 May 1813

Overview
Battle - 4,300 participants

Joint US Army and Navy operation which results in the capture of Ft. George. Americans fail to block the British retreat and the British forces retire to Burlington Heights.

61 Maj. Gen. Winfield Scott circa 1820, National Portrait Gallery

Introduction

Amphibious landings can be tricky as the Americans had learned the previous year when trying to capture Queenston Heights. Hoping to avoid the fiascos of the previous year, US Secretary of War John Armstrong appointed Col. Winfield Scott as Maj. Gen. Dearborn's adjutant general. Dearborn was only too happy when Scott created a general staff and began preparations for the next assault.

Scott's preparations were detailed. British positions were studied and with the help of Oliver Hazard Perry, Scott explored the enemy shoreline. Buoys were placed along the shore of Lake Ontario and along the Niagara River. To ensure the success of this amphibious operation, the US Navy left the safety of Sackett's Harbor and provided the firepower to silence enemy gun positions and suppress any enemy concentrations. The US naval commander on Lake Ontario, Isaac Chauncey, planned the embarkation and troops knew what they were to do and where they needed to go.

The US plan was ambitious. Scott recognized the key to victory was not the reduction or capture of Ft. George, but in destroying the British force across the river. His plan required a successful landing. Once a beachhead was established, troops would continue to advance inland. Having achieved tactical superiority, the next step required that the British Army would not be able to escape. In order to do so, Scott's plan required the US Light Dragoons to play an active role in cutting off the British retreat. To ensure the beach landing went according to his plan, Scott took command of the first wave.

The British commander, Brig. Gen. John Vincent was aware of the American build up, but he did not know where the Americans would land. To this end, Vincent divided his command into three groups, with Ft. George being the center of his defense. Adding to Vincent's problems, the Canadian militia resented the frequent call ups during planting season, especially after the British abandoned them during the American attack on York. On 25 May, the American batteries all along the Niagara River fired on British positions. The following day, heated artillery rounds fired from Ft. Niagara set parts of Ft. George aflame.

The Battle

On the morning of 27th, an American gun fired from Ft. Niagara signaled the start of the assault. As the morning fog dissipated, Chauncey's flotilla, accompanied by 134 boats, each carrying between thirty and fifteen men, began to descend upon the Canadian shore. British batteries which had been earlier identified were silenced and the US squadron moved closer to the landing site. Under the protection of the US squadron, the first wave of Americans "Hit the Beach".

Vincent was aware an American attack was imminent and was at the lighthouse when he saw the American squadron emerge from the fog. To counter the American landing, Vincent sent forth his regular troops to contest the landing. As the first wave landed, Scott was almost bayonetted and lost his footing. Recovering, the Americans pressed forward. As more British troops came up, they pressed the Americans back toward the beach. By this point, the second wave of American troops was landing, and the Americans began to deploy their field artillery. Continuing naval gunfire hampered the British and now facing the landing of the third wave, Vincent chose to withdraw from the field.

Vincent ordered that Ft. George be destroyed and its magazine exploded. As the American army was working its way through the village of Newark, Scott learned of Vincent's intention of destroying the powder magazine. Scott led two companies into Ft. George just as one magazine exploded. Scott was thrown from the captured horse he was riding and broke his collar bone. Undaunted by his injury, Scott was able to save the second magazine and captured the British flag flying over the fort. Scott then continued to pursue Vincent, capturing stragglers and small detachments of retreating British troops.

About five miles south of the fort, Scott ran into Col. Burn and one company of dragoons. Scott wanted to continue attacking the retreating British force but was asked to wait for a second company of US Dragoons to come up. At this point, Scott was ordered by a rival general to stop his pursuit of the retreating British. Scott complied and in doing so, the US failed in its goal of eliminating the entire British force on the Niagara Peninsula.

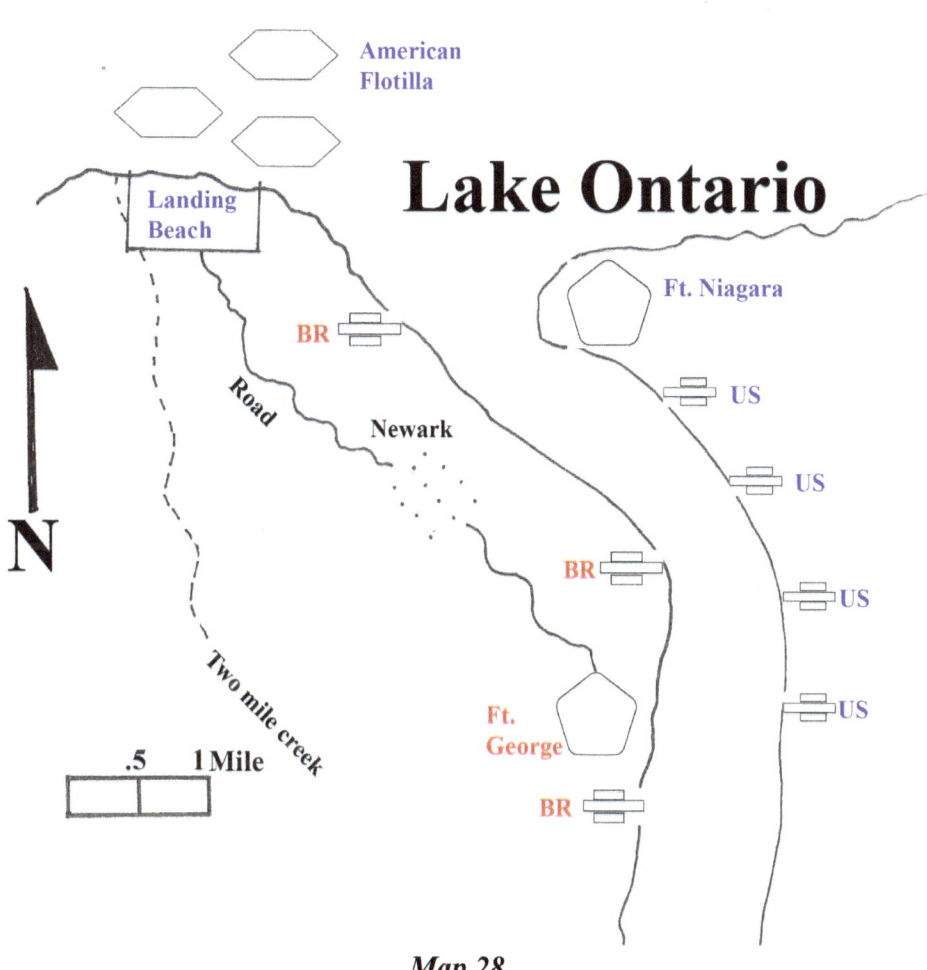

Map 28

Order of Battle

British - 1,134 men
Commanding Officer: Brig. Gen. John Vincent

8th Regiment of Foot, 310 men
49th Regiment of Foot, 232 men
Glengarry Light Infantry Fencibles, 90 men
Royal Newfoundland Regiment of Fencibles, 90 men
Upper Canada Militia, 200 men
Captain Robert Runchey's Company of Colored Men, 27 men
Royal Artillery, 75 men
Provincial Corps of Artificers, 50 men

Grand River Native Warriors, 60 men

US - 4,000 men
Commanding Officers: Maj. Gen. Henry Dearborn, Maj. Gen. Morgan Lewis

Advance Guard, 800 men
Col. Winfield Scott
US Rifles Regiment, one company
Twenty-Second US Infantry Regiment, two companies
Second US Artillery Regiment with one 3-pdr. field gun
Second US Artillery Regiment, one company, serving as infantry
Third US Artillery Regiment, one company, serving as infantry

First Brigade, 1,115 men
Brig. Gen. John P. Boyd
Sixth US Infantry Regiment, 300 men
Fifteenth US Infantry Regiment, 425 men
Sixteenth US Infantry Regiment 270 men
McClure's Volunteers, 120 men

Second Brigade – 1,000 men
Brig. Gen. William H. Winder
Fifth and Twentieth US Infantry Regiments, 250 men
Thirteenth US Infantry Regiment, 400 men
Fourteenth US Infantry Regiment, 350 men

Third Brigade - 1,790 men
Brig. Gen. John Chandler
Ninth US Infantry Regiment, 100 men
Twenty-First US Infantry Regiment, 400 men
Twenty-Third US Infantry Regiment, 350 men
Twenty-Fifth US Infantry Regiment, 650 men
Third US Artillery Regiment, 250 men serving as infantry
US Marines, one company, 40 men

62. American Landing Site, author's photo.

Crossing the Niagara River
US Light Artillery Regiment, two batteries with eight field guns
Second Regiment of Light Dragoons, 2 companies, 150 men

Game Notes

The US clearly learned from its mistakes at Queenston Heights, and the logistical nightmare was not repeated. Scott's meticulous planning and attention to detail, along with the presence of the US squadron, raised the odds for a successful beach landing. Scott's presence in the first wave was critical and he demonstrated both remarkable bravery and a high pain threshold.

That is not to say, however, that the American assault was a forgone conclusion. Many things could have gone wrong, such as the wind on the lake, which could have hindered both naval gunnery and the speedy arrival of American reinforcements. Secondly, though the British troops are outnumbered overall, the British outnumber the Americans during their initial landing. Commanders can be killed, as the high casualty rates among officers

63. Ft George, author's photo.

during the war clearly demonstrates. Dearborn observing from a ship believed that Scott was killed at the landing and could have easily died when he tried to capture Ft. George.

When recreating the entire American assault on Ft. George, or just the beach landing, the unpredictability of the winds on the lake needs to be taken into account. Naval wargames in the age of sail often require a die roll to determine both the direction and the strength of wind. At the beginning of each turn, a D6 is rolled. If a 6 is rolled, then two additional die rolls are required. To determine the strength of the wind, a D6 is rolled and the following modifiers are applied – 1 or a 2, the wind dies down, on a 3 or 4, the wind stays the same, and on a 5 or 6, the wind increases. To determine the wind direction, 2D6s are rolled and the results are applied at the start of the turn: a 2-3, wind turns 90 degrees counter-clockwise, on a 4-5, they turn 45 degrees counter-clockwise, 6 or a 7, they continue to blow in the same direction, on a 8 or 9, they blow 45 degrees clockwise and on a 10 or 11, they blow 90 degrees clockwise.

There are four levels of wind in the game – no wind, slight breeze, slight wind, and strong wind. The game begins with no wind. The wind effects both the speed of the landing craft making it to shore and naval artillery fire. If the wind is blowing against the landing boats, they lose six inches of movement for every level of wind above breeze. If the wind is blowing with the landing craft, they approach the beach much faster and gain six inches of movement for every level of wind above breeze. If the wind is blowing to the side, the boats drift six inches to the east or the west of the landing site. The landing site can only accommodate the craft from each wave, so if the US boats appear before the previous brigade deploys, the men must remain on their craft. The areas around the landing site cannot be used to land troops. It takes two turns for each wave to unload the boats and clear the beach for the next landing.

The wind can also play havoc on the accuracy of the American naval artillery. If the wind is above a slight wind, only half of the naval artillery will be on target. If it is a strong wind, the US naval guns can no longer support the men on the beach.

When refighting the beach landing, the game is limited to ten turns. The game begins with the advance guard hitting the beach and in two turns, the next wave is due to land. However, with the wind, they may not show up at the correct place or time. The British can meet the Americans with two units of their choice, and for every following turn, they can feed in an additional unit.

If one wants to recreate the assault on Ft. George itself, there is a strong chance that the magazine will detonate before Scott can stop it. The fort itself is a ruin, with most of the buildings having been destroyed by heated artillery shells fired from Ft. Niagara the previous day. To simulate Scott's gallantry (or perhaps foolishness), a die roll is required to simulate if Scott and his men are able to stop the three fusses from igniting the magazines.

To recreate this even, the British player writes down on which turn each of the three fuses will detonate. If no American is within 12 inches of a magazine, it explodes with no fanfare. If an American player is within 6 inches of a magazine before it is ready to blow, then the Americans roll a D6 and on anything but a 1, the fuse is cut in time. If any US troops are within 12 to 6 inches of a magazine, the American player will need to roll a 6 on a D6 to stop the explosion – any other resulting die roll results in an explosion.

To calculate the effect of a magazine exploding, the American player needs to roll for effect. In our games we roll one D6 for any figure within 12 inches of a magazine, and on a 5 or a 6, the figure is considered a casualty. If the rule system that you are using tracks wounds and kills separately, determining if any American officer within the blast radius was killed or merely wounded can be determined by a second die roll - with the officer having a 50% chance of dying on the spot.

Victory Conditions

When gaming the entire operation, the American player needs to destroy the British force. For the British player, they need to cause as many American casualties as possible but be able to preserve their armed force by retiring. If playing the entire operation, the British player must survive for twenty-five turns. At this time, Scott, if he is still alive, is ordered to stop his assault on the enemy and the American attack comes to an end.

Further Reading

Barbuto, Richard V., *Staff Ride Handbook for the 1812 Campaigns, 1812-1814*. (Old Fort Niagara Association. Youngstown, NY. 2016)

Cruikshank, Ernest (Lt. Col.), V. D., *The Battle of Fort George*. (Niagara Historical Society. Niagara-on-the-Lake, Ontario. 1990)

Elting, John R., *Amateurs, To Arms!: A Military History of the War of 1812*. (Da Capo Press. New York. 1995)

64. Canadian Fencibles, courtesy of Richard Watts.

Adrian O. Mandzy

Stoney Creek, 6 June 1813

Overview
Battle - 3,500 men

British nighttime attack on an American camp.

Legacy

The War of 1812 forms a critical component in Canadian national identity and the Battle of Stoney Creek is a cornerstone of that idea. Periodically portrayed as "The Battle that Saved Canada" or something similar, the nighttime attack launched against a much larger enemy fits the parameters of 19th and early 20th century ideals of male bravery. Much like the Charge of the Light Brigade during the Crimean War, the attack resulted in high British casualties. The British manage to capture the American cannon, but as they were unable to push the Americans from the field, the British retreated. The leaderless Americans also retreated from the field and upon seeing British warships on their flank, retreated to Ft. George.

Introduction

Following the American capture of Ft. George at the mouth of the Niagara River on 27 May 1813, the British under Brig. Gen. John Vincent began to withdraw their forces toward Burlington Heights. Maj. Gen. Morgan Lewis, who was in tactical command of the US operation, ordered Col. Winfield Scott from pursuing the retreating British. Recognizing his perilous situation, Vincent dismissed the Canadian militia, 500 of which were then captured by the Americans. As the American assault ground to a slow advance, the British began to reinforce themselves at Burlington. From this location, the British could continue to supply Brig. Gen. Henry Procter in western Upper Canada. Given this respite, Vincent was able to re-establish control and keep the retreat from becoming a rout.

65. Stoney Creek Battlefield circa 1860 (Lossing).

To counter the British position at Burlington Heights, the senior US commander Maj. Gen. Dearborn dispatched Brig. Gen. Winder to push the British back. Recognizing the need for reinforcements, Winder asked for more men and Dearborn dispatched a second brigade under Brig. Gen. Chandler. Chandler, being the senior officer, took over command of the operation. Following a small skirmish with Vincent's outpost near Stoney Creek, the Americans set up camp less than five miles away from the British lines. Most of the American troops were encamped in a large field and 800 men guarded the supply boats along the lakeshore.

Having reconnoitered the American position, the British learned that the Americans had tasked their advanced guard with guarding the American camp. Having skirmished with the British earlier that day, the commander of the advanced guard had posted three men as pickets, while the rest went to sleep in a church. Further down the road, the American artillery was camped alongside the thoroughfare. The two brigades were each camped apart from each other and each regiment established their own camp according to their own discretion. What could possibly go wrong for the Americans?

The Battle

Vincent was in a precarious situation. Outnumbered, the most prudent course of action was to withdraw to York, but in doing so, the British supply line to Detroit would be severed. Making a stand at Burlington Heights required surrendering the initiative to the Americans, something Vincent was not wanting to do. Vincent decided to launch a night attack on the Americans.

With unloaded muskets and flints removed from their muskets, 700 British regulars advanced on the American line. The first two American sentries were quickly silenced, but the third American was able to fire a warning before being bayonetted. The British then seized the still sleeping advanced guard.

Moving forward toward the American camp, the British saw the campfires. Presuming they had caught the Americans still sleeping, the British deployed into line and let off a cheer as they charged the American guns. The first charge drove deep into the American line and the gunners scattered after a brief melee. Other American troops began to shoot at the British who quickly began to replace the flints into their locks and began to shoot back.

In the darkness, the senior commanders on both sides were quickly put out of action. Chandler sought to stabilize the line but was captured after being thrown from his horse. Winder too was captured while Vincent managed to get lost during the attack, only to reemerge a few days later. Recognizing the attack had failed to cause an American rout, Lt. Col. Harvey was able to extract the British troops safely back to Burlington, taking with him two captured American artillery pieces. Though the Americans had held the field and caused significant British casualties, the sighting of British warships to their rear brought an end to this advance. Without an American squadron protecting their flank and their supplies, Dearborn ordered the Americans to abandon their advance on Burlington and return to Ft. George.

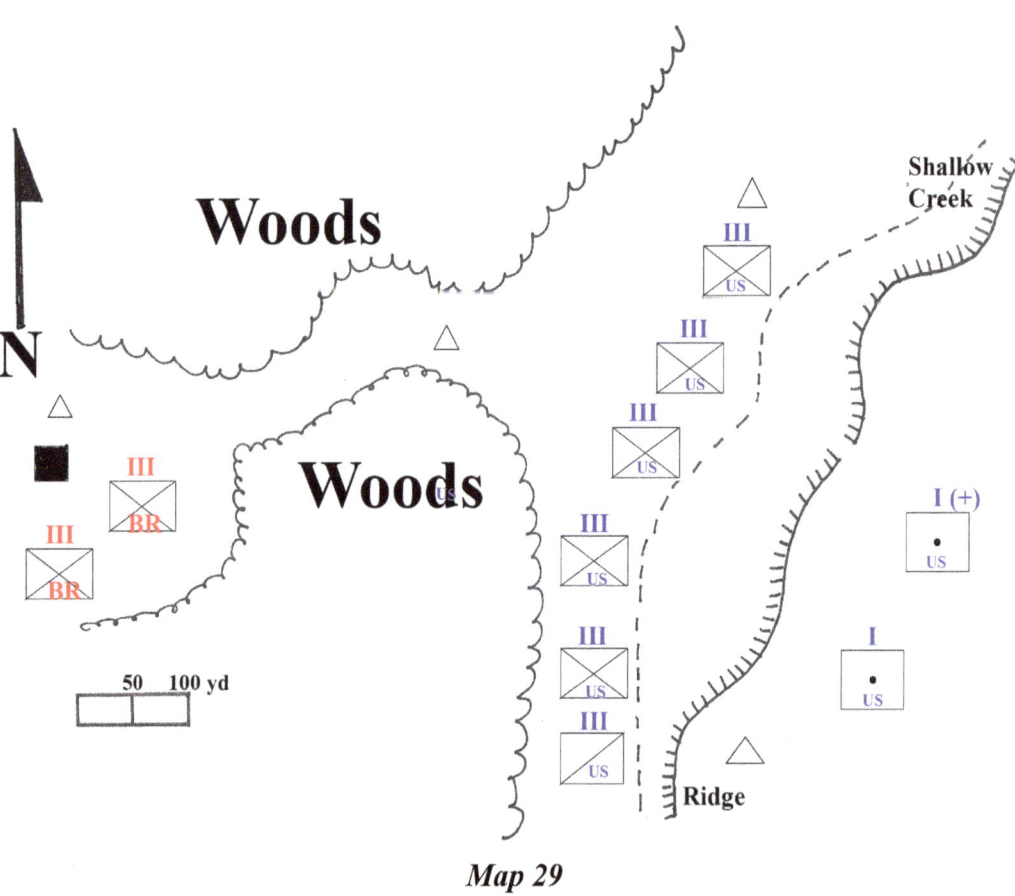

Map 29

Order of Battle

British, 700 men
Commanding Officer: Brig. Gen. John Vincent replaced by Lt. Col. John Harvey

8th Regiment of Foot, 280 men
49th Regiment of Foot, 424 men

US, 2,750 men
Commanding Officer: Brig. Gen. John Chandler

Advance Guard, 100 men
Capt. Jacob Hindman
First US Rifle Regiment, 50 men
Rifle Volunteers, 50 men

Chandler's Brigade, 1,100 men
Ninth US Infantry, 100 men
Twenty-Third US Infantry, 350 men
Twenty-Fifth US Infantry, 650 men

Winder's Brigade, 1,270 men
Fifth/Twentieth US Infantry, 250 men
Thirteenth US Infantry, 400 men
Fourteenth US Infantry, 350 men
Sixteenth US Infantry, 270 men

Division Troops, 260 men
US Light Artillery, 5 guns, 60 men
Second US Artillery, 4 guns, 50 men
Second US Light Dragoons, 150 men

66. US Artillery, courtesy of Roger Chrysler.

Game Notes

A night attack can be difficult to replicate on the tabletop successfully. As noted, the senior commanders on both sides did not have any real command of their troops and in the dark it was hard to distinguish friend from foe. Given the similarities in the cut of the British and American uniforms, soldiers on both sides displayed very similar silhouettes in the shadows of the dying campfires. Friendly fire was a very real occurrence and should be accounted for in any tabletop recreation of a nighttime battle.

The initial British assault on the sleeping riflemen had a far different outcome than their assault on the guns and could be treated as two separate actions. In gaming the attack on the sleeping riflemen, the three pickets first need to be silenced. To see if this occurs, the British force must move toward the church and once they are within 12 inches, the American player rolls three D6. On a roll of a 6, an American picket is able to raise an alarm and the Americans within the church are awakened but will still need a turn to equip themselves. For game purposes, the Americans are considered to be disordered. If the American player is unable to roll a six on the three D6s, the British can try and rush the building without taking any defensive fire.

In gaming the attack on the church, the British will send a hundred regulars of the 49th. Once the Americans are surrounded on three sides, they can be offered terms of surrender. For the British player, they want to

67. Fog of War, courtesy of Mike Vasile.

quickly capture this position and move on. The Americans can try to hold off the British, but once the British secure the entrance and the two other avenues of escape are blocked, they will surrender.

Once the American Riflemen were captured or destroyed, the British player needs to move to the main American camp. Depending on how the British attack on the church goes, the Americans are most likely no longer sleeping. If the American pickets or riflemen begin a firefight with the British, the Americans in camp will rise and form up. If the riflemen continue to hold up the British, the Americans may even move toward the sound of the guns.

It is possible to skip the initial British capture of the riflemen and move to the main assault when the British troops form into line and with leveled bayonets begin their charge. Historically, though the guns were not crewed, and the artillerymen's muskets were neatly stacked in the rear, some of the American infantry was up and under arms to the rear of the guns.

The melee over the guns occurred in the dark and troops continued to pummel each other with whatever was close at hand. The US Dragoons can also be sent into the melee, striking whichever soldier was closer. Generally, the side with the most men involved in a melee wins, but to replicate this nighttime melee, all units participating suffer an additional casualty. The generals who attempted to influence this melee all failed to do so, so in our recreation, they have an exceptional high possibility of being captured or killed. For any general involved in the melee roll a D6. On a roll of anything but a 6, the general is considered to be a casualty.

Though both sides can continue to send in more men into the melee, at some point units will step back and begin shooting at each other. In our game, after the British capture the guns, the American player must send at least three units into the melee, but after that it is the American player's choice if they want to send in more men into the brawl. Rather, the Americans can begin to fire at the British. Given the darkness and smoke that covered the battlefield, any unit that engages in firing may end up firing upon a friendly unit. To replicate this on the tabletop, any unit that fires needs to roll a D6 – on a roll of a 1 or a 2, the unit will end up shooting the closest unit from their side.

Victory Conditions

Historically, once it was clear that the Americans were not put to flight, the British focused on extricating their captured guns from the field. This they did so successfully. Given that the Americans outnumber the British, the British suffered more casualties in the assault. For each captured gun and captured/killed enemy general, the victorious side gains a victory point.

> Extraordinary British Victory - drive the Americans from the field.
> British Victory – hold more guns and captured/killed generals than the Americans
> American Victory – hold more guns and captured/killed generals than the British
> Extraordinary American Victory – completely destroy the British army in front of them

Further Reading

Barbuto, Richard V., *Staff Ride Handbook for the 1812 Campaigns, 1812-1814.* (Old Fort Niagara Association. Youngstown, NY. 2016)

Burton, Pierre, *Flames Across the Border 1813-1814.* (Random House Canada. Toronto. 1981)

Elliott, James E., *Strange Fatality: The Battle of Stoney Creek, 1813.* (Robin Brass Studio. Toronto. 2009)

Elting, John R., *Amateurs, To Arms!: A Military History of the War of 1812.* (Da Capo Press. New York. 1995)

Beaver Dams, 24 June 1813

Overview
Small Battle - 1,500 participants
Small Skirmish - 3 participants

Native Americans and a handful of British troops ambush US troops.

68. Laura Secord, C.W. Jeffrey

Legacy

The Battle of Beaver Dams is enshrined in Canadian popular culture, not as much for the military action, but for the actions of Laura Secord the night before the engagement. In the often-repeated tale, Laura Secord risked life and limb by travelling twenty miles through a swamp in the middle of the night to bring news of an imminent American attack. Her heroism has been the subject of numerous plays, schools have been named in her honor, and her image has been on Canadian stamps and coins. During the Centennial of the War of 1812, a Canadian businessman, Frank Patrick O'Connor, started a chocolate company named for the Canadian heroine. A century later, a stylized portrait of Laura Secord continues to adorn the company logo. Much like the tale of Paul Revere's Midnight Ride, the actions and importance of Laura Secord were undoubtedly much embellished in the 19th century but are included in our tabletop recreation.

Introduction

Following the successful US capture of Ft. George in the spring of 1813, US troops attempted to advance further into Canada, but were stopped at the Battle of Stoney Creek. The Americans then withdrew to Ft. George and the British established themselves at The Forty. To keep an eye on the Americans, the British set up a series of advanced outposts, one of which was at John DeCou's house, a two-story stone building capable of a prolonged defense against any force not provided with artillery. As lines between the two forces began to solidify, British and Natives used the DeCou's house as a base of operations from which they raided American outposts.

Recognizing the importance of DeCou's house, the Americans formulated a plan to neutralize it. A strong American force under Col. Boerstler and Maj. Chapin was assembled at Ft. George. On their way to DeCou's house, they stopped for the night in Queenston. It is here that Laura Secord, the wife of James Secord, overheard the American plans and as her husband was still recovering from wounds received earlier during the Battle of Queenston Heights, she set off to warn the British. Travelling on back wood paths, Secord stumbled upon a Native American camp and was brought before Lt. FitzGibbon, where she gave him the news.

Meanwhile, the American forces began their advance westward. The force was spotted by a group of Native Americans and a brief exchange of gunfire occurred. More Native Americans appeared on the American flanks. The Americans pushed their cannon and wagons up the Niagara escarpment and upon reaching the crest, they rested. The sounds of gunfire had given plenty of warning and local farmers climbed onto their roofs to watch the upcoming battle. As the Americans continued their advance, the road they travelled entered a well-timbered track of pine and Beachwood trees. The Canadians were aware of a gully that crossed the road in the wooded track and picked it as their place of ambush.

The Battle

Watching the American advance, the Native Americans took up positions on both sides of the road in the thickly wooded area, with the British line being held in reserve. An advanced group of Chapin's men rode in front of the column and they were the first to be attacked. Native Americans fired a volley, which struck nineteen of the twenty men down. During this chaos, the Americans were attacked on both their flanks. A small number of Native Americans also struck the rear of the column, which caused a stampede of men and horses into the middle of the column.

The Americans survived the initial attack and were successful in establishing a defensive perimeter. With a disciplined bayonet charge, Col. Boerstler pushed his men through the woods into the surrounding fields. Deploying his artillery, an opportune canister shot caused a number of Native American casualties. At this point, a number of Mohawks left the battle, while others carried on.

Observing that the Native American fire was decreasing, Col. Boerstler decided to continue his advance on the DeCou's house. Recognizing that the wagons were slowing his advance, Boerstler split his command. Leaving his guns, wagons and cavalry behind in the relative safety of the wooded gully, Col. Boerstler formed his infantry and began to march down the road. Fearful of another Native American ambush, Col. Boerstler deployed a company on each flank to cover his main marching column.

The American column barely began its forward movement when a large group of Native Americans assaulted the Americans from two sides. Boerstler had a horse shot under him early in the fight and suffered a second wound in his thigh. Unable to move forward and taking fire on his right flank, Boerstler began to withdraw toward his wagons and cannon. Having exhausted most of the ammunition carried by the infantry, the return to the wagons allowed the men to resupply themselves. Pinned down and unable to move forward, the Americans began to contemplate their breakout.

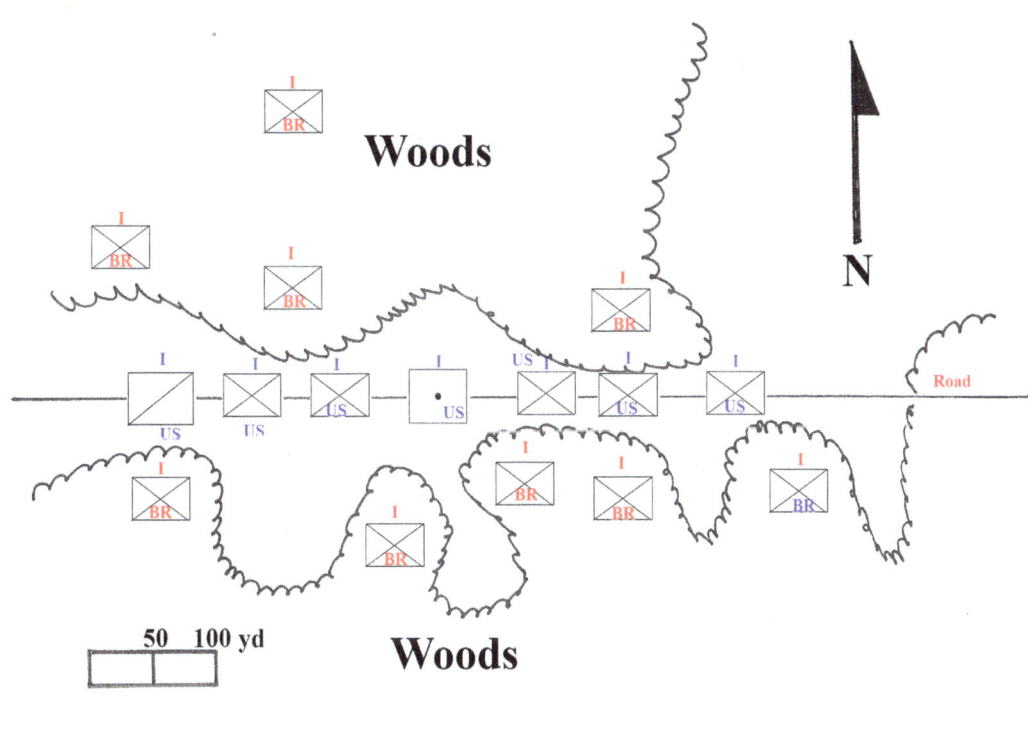

Map 30

With the arrival of fresh British troops on to the scene, Lt. FitzGibbon approached the Americans to discuss their surrender. The appearance of Capt. John Hall at the head of a troop of newly uniformed Provincial Dragoons, along with the No. 2 Company of the Lincoln Militia on the battlefield, gave credence to FitzGibbon's claim that the Americans were outnumbered. Having lost a number of men, including a number of officers, in the previous three hours of fighting, the exhausted Americans surrendered.

Order of Battle

British - 450 men
Commanding Officer: Lt. James FitzGibbon

Kahnawake Mohawks with Capt. Dominique Ducharme of the Indian Department, 300 warriors These should be divided into 6 warbands.

Mohawks with Capt. Johnson Kerr, 100 warriors. Two or three warbands.
49th Infantry Regiment, 46 men – held in reserve
Reinforcements
Provincial Dragoons, one troop, *20 men*
No. 2 Company of the Lincoln Militia, *60 men*

US - 1,000 men
Commanding Officer: Col. Charles G. Boerstler, Fourteenth US Infantry Regiment

Fourteenth US Infantry Regiment, 600 men
Sixth US Infantry Regiment, one company, *60 men*
Thirteenth US Infantry Regiment, one company, 60 men
Twenty-Third US Infantry Regiment, one company, *60 men*
US Dragoons, 20 men
Artillery Battery – one 6-pdr. gun, one 12-pdr. gun, two caissons (ammunition carts)
Four Supply Wagons

Maj. Cyrenius Chapin, New York State Militia Mounted Volunteers, 50 men

69. Native Americans, courtesy of Richard Watts.

Game Notes

The battle is strictly a fight between American troops and Native Americans. No British troops took part in the actual battle, and they only appeared at the end to negotiate the American surrender.

We have gamed this battle on a number of occasions with different outcomes. When trying to recreate the battle, the American player is in a difficult position. The Native Americans know the lay of the land and picked the perfect spot to set the ambush. Once the Americans are able to recover from the initial shock, they must go on the offensive. Native Americans will not be able to take heavy casualties and are brittle. Bayonet charges and well positioned artillery, if used successfully, can drive the British troops away. To win the game, the American player must survive the ambush and get their troops moving in good order toward the DeCou house.

For the British player who is used to commanding loyal redcoats, Native Americans were a product of a different society and thus should behave differently on the tabletop. Native Americans do not follow orders blindly, as they have different priorities and goals from the British officers with them. By no means are the British officers controlling the warriors. Both brave and bold, the warriors acted independently in their own interests – be it loot or honor. They excelled in the tactics of feigns, fake retreats, ambushes and counter ambushes but were unlikely to stay around in a pitched battle, especially if there was no apparent victory. Unlike regular troops who could readily replace their casualties from overseas, each Native American casualty was a member of the greater community who was irreplaceable. Thus, if a group of Native Americans suffer casualties, they may simply leave the battlefield, much to the chagrin of the British officers with them.

Victory Conditions

To win the game, the American player must survive the ambush and get their troops moving in good order toward the DeCou house. If the Americans suffer more than forty percent casualties before reaching the DeCou house they achieve a minor victory and if they suffer less than forty percent before reaching the DeCou house they gain a major victory. If the Americans are unable to reach the house they lose the game.

Small Skirmish - Gaming the night travels of Laura Secord

For those wanting to include the actions of early feminist heroine Laura Secord in their game, it is possible to do so. Few figures are required, as the heroine has only her stealth and wit to move through the hostile landscape. Besides the figure of Secord, we use a few animals, such as bears, deer and wild geese. As the actions of Laura Secord are the only ones that matter, the recreation is well suited for solo gaming. Place a figure of Laura Secord at one end of the table and a Native American encampment at the other. Then randomly scatter some terrain (we used 3 woods, 2 farm fields with fences, and 2 swamps) between the two objectives. For accurate placement, the player closes their eyes and lightly throws the terrain in the direction of table. If a terrain piece falls off of the table, re-throw it. This recreation lasts for 12 turns on a 4 foot by 6 foot table.

Laura can move at three speeds – carefully (2 D6), briskly (4 D6) or running (6 D6). At the start of the game, Laura states which speed she wants to move at. If she moves at careful speed, just move the figure the result of the two die rolls. If moving at faster speeds, there is a chance she may fall down. When a total of 3 sixes appear in the die roll, Laura falls down and must spend the following turn getting up. The following turn she can only move at careful speed (2 dice). At this point she needs to roll a D6 and on a roll of a 1, she becomes disoriented and moves for that one turn in the opposite direction.

If our heroine encounters any fields or swamps, her movement is reduced by her highest single die roll. Thus, if she was moving briskly and rolled a 3, two 4s, and a 5, she would move 11 inches. The previous days of rain softened the ground considerably, which reduced movement across plowed fields. If Laura crosses a farm field, her movement is reduced by her lowest single die roll. Thus, if she was moving briskly across a field and rolled a 3, two 4s, and a 5, she would move a total of 13 inches.

By 1813, bears were all but hunted out of existence in the Niagara region, but we include the possibility of our heroine running into one. At any time, if Laura moves through a forest, she needs to roll an additional D6. If a 1 is rolled on a D6, she encounters a bear. A second D6 is then rolled. On a roll of a 2-6, she is able to run away from the fury of a bear and for the next three turns needs to move at running speed. If she rolls a 1, the bear eats Laura. The geese and deer play no part in the game but certainly the presence of the animals on the table adds to its visual appeal.

To add more drama to the event, Laura may run into a pair of Americans scouting the area for food and loot. For every turn Laura is running, an additional D6 needs to be rolled at the end of the turn. If the player rolls a 1, then a second die roll is needed to determine where the two scouts will appear – on a 1, the men are behind her, on a 6 they are in front of her, on a 2 or 3 on her right and on a 4 or 5 on her left. If Laura runs again, the two men will hear her and move towards her at their maximum speed to try to capture her.

70. Fourteenth Infantry being flanked by Native Americans, courtesy of Roger Chrysler.

Further Reading

Benn, Carl, *The Iroquois in the War of 1812.* (University of Toronto Press. Toronto. 1998)

Dewar, Donald Keith, *The Battle of Beaverdams: The Story of Thorold's Battle in the War of 1812.* (Slabtown Press. St. Catharines. 1996)

Adrian O. Mandzy

Raid of Ft. Schlosser, 5 July 1813

Overview
Small Skirmish - less than 100 participants

Successful British raid on an American supply depot.

Introduction
During the mid-18th century, the French had built a small military fort to guard the Portage Road around Niagara Falls. Destroyed and rebuilt by the British, the palisaded four bastion earthwork was in disrepair at the start of the War of 1812. A small US garrison guarded the fort, and its buildings were used to store supplies.

The Battle
In the early morning of 5 July, a small British force crossed the Niagara River in three boats. Landing undetected, the British were able to surprise the garrison and captured a number of supplies. Remaining on site for less than half an hour, the British began to return home when a group of American troops appeared. The Americans gave chase, but the British were able to return home with their loot.

Though the British burned the fort, along with many other buildings on the American side of the river in the winter of 1813/4, the earthen walled enclosure continued to serve the American Army as a supply base through the following summer.

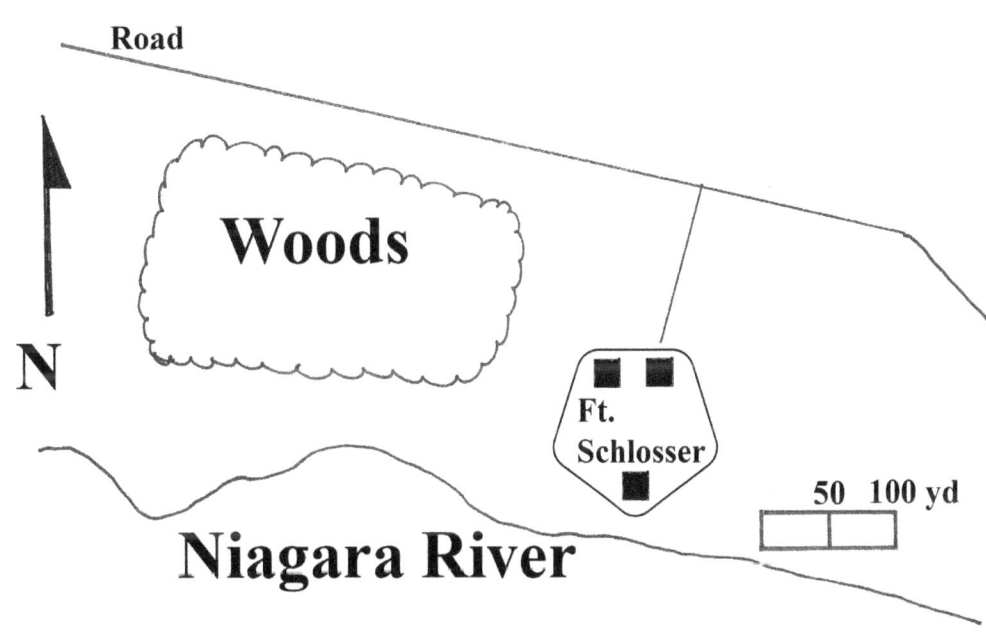

Map 31

Order of Battle
British - 34 men
Commanding Officer: Lt. Col. Thomas Clark

49th Regiment of Foot, 6 men
Militia, 20 men

US - 36 men
Commanding Officer: Unnamed Lt.

At the fort
    Unnamed Lt.
    Wagon Master Palmer
    New York State Militia, 8 men

Armed civilians, 6 men
One 6-pdr. cannon
Reinforcements
20 New York State Militia

Game Notes

This small-scale engagement, which is typical of the ongoing petit war that existed all along the frontier. Given the scale of the forces involved, one could game the raid at one-to-one. When gaming this scenario, the earthworks should be considered to be run down and individuals could most likely move through the old stockade but would need to use the fort's gate to remove barrels of food or ammunition.

In out recreation, the Americans have ten supply counters scattered throughout the fort's two storehouses, barracks and officer headquarters. The British should be able to land on the American side of the Niagara River but as they approach the fort itself, it is possible an American would spot them. If the British are spotted, then the Americans can try to stop their assault – either at the fort's walls or in each individual building. The 6-pdr. field gun can be fired but is considered to be unloaded at the start of the game. Ammunition for the cannon is next to the piece.

At the end of the fifth turn, American reinforcements appear at the eastern table edge. The game plays for ten turns.

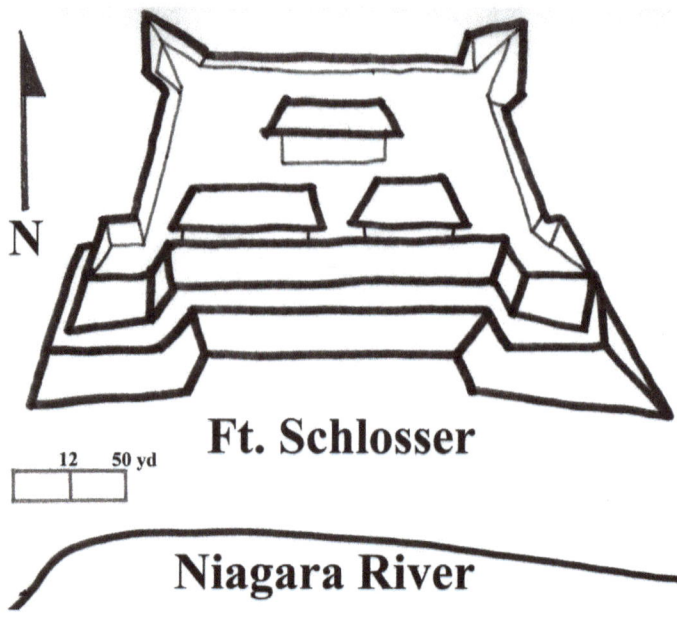

71. Plan of Ft Schlosser, drawn by Christian Wright.

Victory Conditions

The British gain three victory points if they capture the cannon and are able to wheel it to their boats and acquire two victory points for each supply counter moved to their craft. Each supply counter remaining in American hands or under British control but not at the river's edge by the end of the game grants the owner each supply counter one additional supply point. Burning the buildings or the fort does not help nor hurt the British player but will cause them to spend a turn trying to start a blaze, time their troops should be doing something else.

72. New York State Militia, courtesy of Richard Watts.

Further Reading
Collins, Gilbert, *Guidebook to the Historical Sites of the War of 1812.* (Dundurn Press. Toronto. 2004)
Cruikshank, E. (ed.), *The Documentary History of the Campaign upon the Niagara Frontier in the Year 1813.* (Lundy's Lane Historical Society. Welland, Ontario. 1903)
Lossing, Benson J., *Pictorial Field-Book of the War of 1812.* (Harper& Brothers Publishing. New York. 1869)
Wilder, Patrick, *Seaway Trail Guidebook to the War of 1812.* (Seaway Trail, Inc. Oswego, NY. 1987)

Adrian O. Mandzy

Black Rock, 11 July 1813

Overview
Small Skirmish - less than 500 participants

British troops cross the Niagara River and capture the American settlement of Black Rock. The Americans launch a counterattack and the British suffer a number of casualties while withdrawing.

Introduction
Following the success of the British raid on Ft. Schlosser, Sir George Prevost ordered Lt. Col. Bisshopp to plan a raid on Black Rock. Founded in 1802, the village of Black Rock was an economic rival of Buffalo, located just four miles away. Named for the large black limestone rock that stood in the Niagara River, the village benefited from the naturally protected harbor formed by the rock. Where the nearby Conjocta Creek emptied into the Niagara River, a shipyard built boats for the American navy on Lake Erie. By the War of 1812, the village's warehouses were protected by a blockhouse, an earthen battery and a small garrison.

The Battle
At 0200, the British boarded their boats and crossed the Niagara River. Landing closer to Black Rock, FitzGibbon advanced on the Americans under a flag of truce. Meeting with the American commander Maj. Hall, FitzGibbon called on the militia to surrender and allowed them to retire on parole. Before FitzGibbon completed his speech, the militia within earshot broke ranks and began fleeing towards Buffalo. FitzGibbon then ordered Hall to stop his men from retreating, but Hall replied he could not and ran off after his command.

New York State Militia Brig. Gen. Peter Porter, who resided at his house at Black Rock, was awoken by the British landing and attempted to reach the American camp. Learning of the militia's rout, Porter traveled on foot to Buffalo. Meeting Capt. Cummings and a hundred of his US regulars, who were marching to the sound of gunfire, Porter ordered Cummings to halt. Taking a horse, Porter rode to Buffalo and recruited the remains of Adam's Militia, militia from Buffalo itself, and a group of Seneca warriors.

At this point, Bisshopp's men had landed and began to destroy public property. The blockhouse, barracks and navy buildings were put to torch. At this point, accounts differ as to what happened next. The British sources state that had taken all that they could and that they were ready to depart from the American side of the Niagara River, while the American sources state the British were still occupying the village. In any event, at some point Porter launched a coordinated attack on the British and was able to inflict a number of casualties. In the course of a running firefight, the British retreated to their boats and crossed the river.

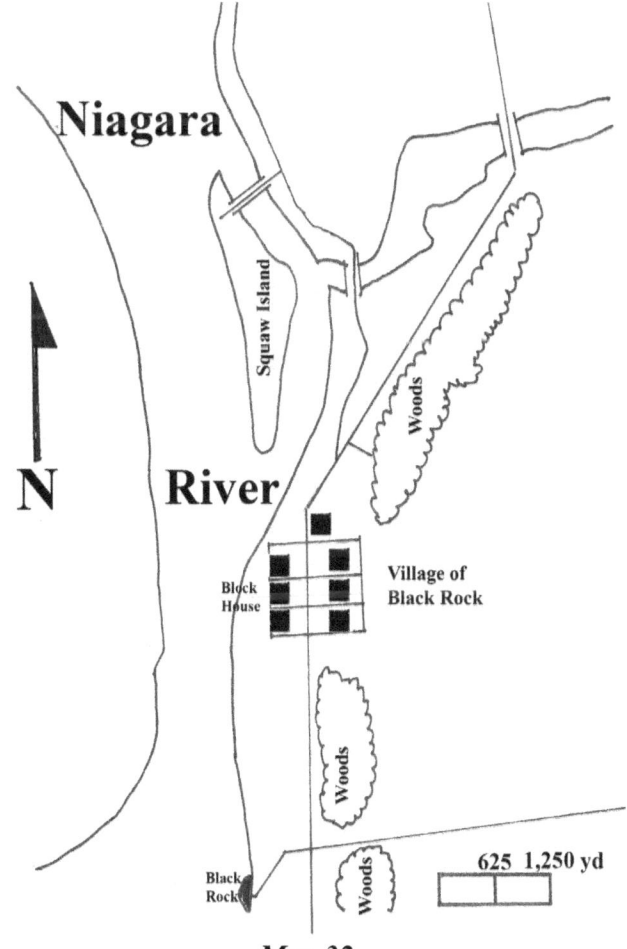

Map 32

Order of Battle
British - 250 men

Commanding Officers: Lt. Col. Cecil Bishop
   Lt. James FitzGibbon

8th Regiment, 40 men
41st Regiment, 100 men
49th Regiment, 40 men
Lincoln Militia, 2nd and 3rd Regiment, 40 men
Royal Artillery, 20 men

US
Initial force - 150 men
Maj. Parmenio Adams
New York Militia, 150 men

Reinforcements - 300 men
Brig. Gen. Peter Porter
US Regulars, Capt. Cummings, 100 men
Buffalo Militia, 50 men
Rallied New York State Militia, 100 men
Seneca Warriors, 40 men

73. Gen. Peter Porter, (Lossing).

Game Notes

There are various ways of recreating this historical event on the tabletop. One can duplicate the entire raid, starting with the British landing. It is possible that the American militia may not flee following FitzGibbon's speech and attempt to resist the British. The presence of American cannons and a blockhouse could work to slow down the British advance.

In such a game, the American regulars under Cummings would appear at the start of turn six, the Seneca on turn eight, and the Buffalo Militia on turn ten. Such a game would play for twelve turns.

Victory points in such a game would be given to the British for the destruction of public property – the two warehouses, the blockhouse, the barracks building, and an American sailing vessel are worth one victory point apiece. In addition, the British would gain a victory point for each of the six government supplies scattered around Black Rock that they capture and bring to their boats. If a unit is attempting to set fire to a building or move a supply counter, they cannot do anything else that turn.

The American focus is to destroy the British force. While preventing the destruction of property is important, the Americans only gain victory points for causing British casualties. Thus, for every two rank and file casualties the Americans cause, they gain one victory point. Two victory points are given for the death of a British officer.

The other option for gaming this scenario is to focus on the American counterattack. FitzGibbon warned Bisshopp that the Americans may come back and that they needed to move quickly. By this time, the physical property was already destroyed, so victory points can only be earned by either killing the British or by taking the ten scattered captured American bounty markers back across the river. Units moving captured stores move at half their movement rate. A unit must spend a full turn loading a supply marker into a boat before they then row into the river. Units rowing can be fired upon for one turn while they are in the river.

Placement of the British troops and the scattered stores requires a random placement. Looking at this scenario map, note the series of numbers along the map edge – six numbers along the short side of the map and ten numbers along the long side of the map. A roll of two dice, a D10 and a D6 provides a coordinating spot for the placement of each pile of loot. Once all of the bounty is placed on the board, the British player would deploy their troops as they see fit.

Porter launched a coordinated attack with the combined force. In this tabletop recreation, once the British player places their troops, the Americans then can bring their troops on from one of the three land-based table edges. Boats in the river can be shot at for one turn, but the boat itself is considered to be soft cover. A lucky artillery shot could sink a boat.

Victory Conditions
The side with the most victory points wins.

Further Reading
Cruikshank, E. (ed.), *The Documentary History of the Campaign upon the Niagara Frontier in the Year 1813.* (Lundy's Lane Historical Society. Welland, Ontario. 1903)

Fitzgibbon, Mary Agnes, *A Veteran of 1812: The Life of James Fitzgibbon.* (Coles Publishing Company. Toronto. 1979)

Lossing, Benson J., *Pictorial Field-Book of the War of 1812.* (Harper & Brothers Publishing. New York. 1869)

74. American militia and Native allies, author's photo.

75. 8th Foot, courtesy of Knuckleduster Miniatures.

Burlington Races, 28 September 1813

Overview
Naval Engagement

The American and British battle on Lake Ontario.

Introduction
    In June 1813, the American Navy on Lake Ontario cautiously ventured from the safety of Sackets Harbor. The launch of the *General Pike* gave the Americans more fire power than the two largest warships in Yeo's flotilla, but Chauncey was hesitant to risk damaging his squadron. An aborted combined Army/Navy raid on Burlington Heights resulted in a second raid on York, which brought about the release of a few American prisoners and the burning of public property.
    On 7 August, Yeo's flotilla made an appearance on the lake and the two squadrons observed each other at a distance. A sudden lake squall caught both fleets and the two largest top-heavy American schooners, the *Hamilton* and the *Scourge*, capsized. Two days later, the small American schooners *Julia* and the *Growler* fell behind and were captured by the British. In the span of three days, Chauncey managed to lose four vessels without even firing a shot.
    With the commissioning of the *Sylph*, Chauncey ventured out into the lake once more. On 11 September, the Americans caught sight of the British on the open lake. Following Perry's total victory over the forces of the Crown on Lake Erie, Chauncey went against his instincts and chose to attack the Royal Navy near the mouth of the Genesee River. For more than three hours, the two flotillas battled without a decisive outcome. With the wind picking up, Chauncey's flotilla managed to escape. Two weeks later, Chauncey spotted Yeo's squadron in the western part of the lake.

The Battle
    With the wind at his back, Chauncey's squadron began to move on the British flotilla. Knowing that some of his vessels were top heavy, Chauncey tried to stabilize them by having them towed by his larger craft. This not only slowed Chauncey's squadron, but reduced both the speed and maneuverability.
    Yeo recognized that the British needed to keep their flotilla, for without it Upper Canada would be lost. Yeo turned his fleet around and began to make for the shelter of Burlington Bay. During the opening salvoes, *General Pike's* long guns damaged (some say crippled) the *Wolfe*. Capt. William Mulcaster of the *Royal George* brought his sloop around to cover the *Wolfe* but was in turned battered by the *Madison* and the *Governor Tompkins*. Chauncey pursued, but as he was unwilling to cut loose his smaller top-heavy schooners, the Americans moved slowly.
    Yeo was able to get his squadron in a line just offshore and waited for the American assault. The *General Pike* had suffered damage in the opening salvos when one of its cannons exploded on the forward gun deck. Concerned with the weather, Chauncey blinked and withdrew from the fight.

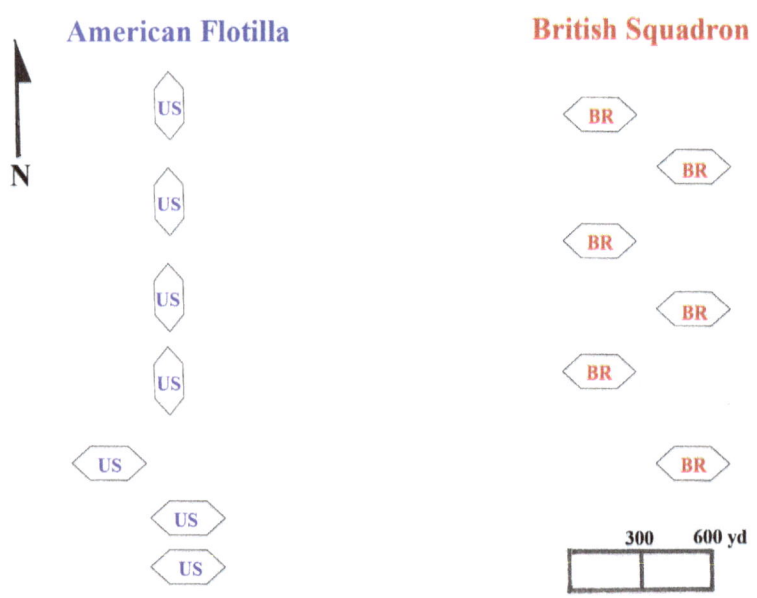

Map 33

Order of Battle
British
Commanding Officer: Commodore Sir James Lucas Yeo

HMS *Wolfe*, frigate, one 24-pdr., eight 18-pdr., four 68-pdr. carronades, ten 32-pdr. carronades

HMS *Royal George*, frigate, three 18-pdr., two 68-pdr. carronades, sixteen 32-pdr. carronades

HMS *Lord Melville*, brig, two 18-pdr guns, twelve 32-pdr. carronades

HMS *Earl of Moira*, brig, two 9-pdr. guns, twelve 24-pdr. carronades

HMS *Beresford*, schooner, two 12-pdr., ten 32-pdr. carronades

HMS *Sir Sidney Smith*, schooner, one 12-pdr., one 9-pdr., six 18-pdr. carronades

US
Commanding Officer: Commodore Isaac Chauncey

USS *General Pike*, frigate, twenty-eight long 24-pdr. guns
USS *Madison*, frigate, twenty-eight 32-pdr. carronades
USS *Sylph*, brig, four 32-pdr guns, six 6-pdr. guns
USS *Oneida*, brig, sixteen twenty-four 24-pdr. guns
USS *Governor Tompkins*, schooner, one 32-pdr., one 12-pdr., four 6-pdr. guns
USS *Conquest*, schooner, one 32-pdr., one 12-pdr., four 6-pdr. guns
USS *Asp*, schooner, one 24-pdr.

76. Commodore Isaac Chauncey, USN by Gilbert Stuart circa 1818 US Naval Academy Museum Collection.

Some sources list the - USS *Growler*, schooner, one 32-pdr., one 12-pdr. – but she is on the British side until her recapture in Oct 13.

Game Notes

Yeo's purpose is to stop the Americans from gaining control of Lake Ontario and to do so, he must preserve his fleet. Chauncey, on the other hand, must destroy the Royal Navy on Lake Ontario for the Americans to advance into Canada.

It has been argued that any other US naval commander would have gone for the win. Chauncey chose not to go for the glory and thus continued to do what he did best – build ships. Like Gen. George B. McClellan fifty years later, Chauncey excelled at building up a force, but was unwilling to use it.

Both of the naval encounters fought in September 1813 on Lake Ontario were indecisive due to Chauncey's actions, or perhaps more accurately, a fear of commitment. Both engagements, however, can easily be recreated on the wargaming table.

The engagement fought on 11 September near the mouth of the Genesee River occurred in open water with no visible shoreline. The Americans have the wind, but the wind changes during the course of the battle and the British were able to withdraw. In our recreation, there are four levels of wind in the game – no wind, slight breeze, slight wind, and strong wind. The game begins with a slight breeze pushing the Americans forward.

Each naval gaming system should have a rule for wind. On Lake Ontario, the wind is known for being erratic and can change both speed and direction rather quickly. To replicate this instability, a D6 is rolled at the beginning of each turn. If a 6 is rolled, then two additional die rolls are required. To determine the strength of the wind, a D6 is rolled, and the following modifiers are applied – 1 or a 2, the wind dies down, on a 3 or 4, the wind stays the same, and on a 5 or 6, the wind increases. To determine the wind direction, 2D6s are rolled and the results are ap-

77. USS General Pike by Charles Ware circa 1820, National Archives.

plied at the start of the turn: a 2-3, wind turns 90 degrees counter-clockwise, on a 4-5, they turn 45 degrees counter-clockwise, 6 or a 7, they continue to blow in the same direction, on an 8 or 9, they blow 45 degrees clockwise and on a 10 or 11, they blow 90 degrees clockwise.

The second engagement was fought on the 28th. One can start the game with the British flotilla deployed in line and with both squadrons having suffered damage from earlier in the day, or the recreation can begin at the start of the day.

To recreate the engagement at the start of the day, the *Wolfe* should be placed 24 inches northwest of the southeast corner of the table, with the rest of the Royal Navy behind her. The British flotilla should be facing northeast. The *General Pike* should be placed 36 inches northwest of the *Wolfe*, facing southeast. The *General Pike*, *Madison*, and *Sylph* should be towing the *Governor Tompkins*, *Conquest*, and the *Asp* behind them. To release a schooner from a tow, both ships must spend a turn to cast off. If there is a strong wind, the American player needs to roll 2D6. If a 2 is rolled, the schooner is swamped and sinks. This recreation plays on a 6 foot by 4 foot table.

If one wanted to game the American assault on the British line, a die roll would be required to determine the level of damage the *Wolfe, General Pike,* and the *Royal George* have suffered at the onset of the engagement. If using Warlord's *Black Seas Rules*, consult the Critical Table - Hull and the Critical Table – High to determine the damage to each craft. Roll two D6 to determine the amount of damage each ship has suffered from the initial meeting.

Wargaming an attack by the American flotilla on the British line also makes for a very manageable solo game. In such an action, the player would set up the British boats in a line in the middle of the table. Once the player finishes placing the Royal Navy boats, the player then takes command of the American squadron and attacks the forces of the Crown. Ship to ship combat would take place as usual, with the player rolling for both sides. Once a British craft is sunk or strikes its colors, the remaining boats of the Royal Navy can attempt to flee. The player would then begin to move the remaining British craft to the northwest corner of the gaming table at all possible speed. If the British manage to extract four boats off the northwest corner of the board, the Americans automatically lose the game.

Victory Conditions

Yeo's purpose is to stop the Americans from gaining control of Lake Ontario and to do so, he must preserve his fleet. Chauncey, on the other hand, must destroy the Royal Navy on Lake Ontario for the Americans to advance into Canada. Thus, the Americans must sweep the British from the Lake. Anything less than the complete destruction of the Royal Navy results in an American defeat.

The game plays until the Royal Navy is sunk or manages to successfully extradite itself from the area.

Further Reading

Elting, John R., *Amateurs, To Arms!: A Military History of the War of 1812.* (Da Capo Press. New York. 1995)

Lardas, Mark, *Great Lakes Warships 1812-1815*. (Osprey Publishing. Oxford. 2012)

Lossing, Benson J., *Pictorial Field-Book of the War of 1812*. (Harper& Brothers Publishing. New York. 1869)

Malcomson, Robert, *Lords of the Lake: The Naval War on Lake Ontario, 1812-1814*. (Naval Institute Press. Annapolis. 1989)

Williamson, Robert J.,The Burlington Races Revisited: A Revised Analysis of an 1813 Naval Battle for Supremacy on Lake Ontario. *Canadian Military History*. Vol. 8, No. 2. 7-15. 1999

Chippawa, 5 July 1814

Overview
Major Battle - 5,000 participants

An American and a British brigade fight over a flat meadow and both sides suffer heavy casualties.

Legacy
While most Americans have no idea about the War of 1812, the Battle of Chippawa continues to be commemorated in the US Army. According to an unsubstantiated legend, the gray uniforms worn by US Army Cadets at West Point, as well as other military institutions, were chosen to commemorate the men of Scott's brigade. The often-repeated phrase, "Those are Regulars, by God!" was used to help inspire cadets and provided a positive legacy for the US Army.

Introduction
With the defeat of Napoleon at Leipzig in the fall of 1813, the balance of power had shifted in Europe. By April 1814 Napoleon had abdicated, which left the British with the manpower and finances to spend on North America. Recognizing that Chauncey was unwilling to launch an attack on Kingston, much less risk a single ship when the British attacked Oswego to deprive the US Navy of much needed supplies, confused orders from US Secretary of War John Armstrong led Maj. Gen. Jacob Brown to invade the Niagara Peninsula for a third year in a row. The goal of the US was to capture Burlington Heights and in doing so, the British would be hard pressed to supply their forces in the west.

Crossing the Niagara River in two places, the Americans found themselves surrounding Ft. Erie. After a few shots, the lightly garrisoned Ft. Erie surrendered. The Americans continued to press northward along the Niagara River. The British had fortified the bridge crossing of the Chippawa River with blockhouses, barracks and an earth work called a *tete-de-point*, which was a two-sided triangle. Upon arriving at Chippawa, Brig. Gen. Scott observed that more troops were needed to force the crossing and withdrew to Street's Creek, where the Americans set up camp. The following day, Scott was joined by New York State Brig. Gen. Porter's Third Brigade.

78. Sir Phineas Riall, courtesy of John R. Lee.

The Battle
As the Americans continued their reconnaissance, Native Americans loyal to the Crown, along with Canadian militia, began filtering through the woods toward the American lines. Using the woods as cover, these British troops began sniping at the Americans. Brown then ordered Porter's brigade to clear the woods. Porter's men, which also included a number of Iroquois warriors, were successful in driving the skirmishers back to the Chippawa. The British then committed their light troops, which pushed Porter's men back to Street's Creek.

The British commander, believing that he was facing only "Buffalo Militia", left his defensive works and strove to drive the impotent Americans from the field. At the same time, Scott was preparing to drill his men on the relatively flat open meadow between the two camps. Warned by Porter of the coming British advance, Scott ordered an artillery company to join his maneuvers.

The British had already taken the meadows and when the Americans began to deploy into line, the British artillery began to fire. Seeing the gray coated men deploying in the face of artillery fire, the British commander Riall famously stated "Those are Regulars, by God!" To secure the American left flank, Scott sent the Twenty-Fifth Regiment into the woods.

In the ensuing artillery duel, the Americans scored a hit on a British caisson, which promptly exploded and put the remaining British guns out of action. Amidst the artillery duel, both the British and Americans formed up for action. The Americans deployed into line, while the British came on in regimental columns. As the British advanced, Scott adjusted his line and formed a very shallow V-shape. By this point the Twenty-Fifth had driven back the British lights and appeared on the British flank. The American line fired a number of times and the British advance stalled. At this point, Scott ordered a bayonet charge, which broke the British advance. As the British began to retire to their defensive line, a second British three gun 6-pdr. battery was brought into action and deployed against the Americans. As the British reached the safety of their prepared positions, Scott was able to halt his troops. Though Brown wanted to attack the British defense works directly, he was persuaded by his officers to follow his original plan, which was to force a crossing of the Chippawa River further west. Skirmishing continued until nightfall, when both sides retired to their respective camps.

Map 34

Order of Battle
British - 2,170 men
Commanding Officer: Maj. Gen. Phineas Riall

1st Regiment of Foot, the Royal Scots, 500 men
8th Regiment of Foot, The King's Regiment, 480 men
100th Regiment of Foot, 450 men
Royal Artillery, 70 men
Lincoln Militia, 300 men
Native Warriors, 300 men

US - 2,500 men
Commanding Officer: Maj. Gen. Jacob Brown

First Brigade, 1,319 men
Brig. Gen. Winfield Scott
Ninth and Twenty-Second US Infantry Regiments, 549 men
Eleventh US Infantry Regiment, 416 men
Twenty-Fifth US Infantry Regiment, 354 men

Third Brigade, 936 men
Brig. Gen. Peter B. Porter
Fifth Pennsylvania Regiment (Fenton's Pennsylvanians), 500 men
Seneca and Cayuga Warriors, 386 men

79. Scott's Brigade, courtesy of Knuckleduster Miniatures.

Canadian Volunteers, 50 men

Corps of Artillery, 200 men, Maj. Jacob Hindman
Capt. Thomas Biddle, three 12-pdr. guns
Capt. John Ritchie, two 6-pdr. guns and one 5.5-inch howitzer
Capt. Nathan Towson, two 6-pdr. guns and one 5.5-inch howitzer
Capt. Alexander Williams, three 18-pdr. guns
Lt. David Douglas's Company of Sappers, Bombardiers and Miners, two 18-pdr. guns

Dragoons, 70 men
US Light Dragoons, one company
New York Volunteer Dragoons, one company

Game Notes

Though the Battle of Chippawa proved to be strategically indecisive, it remains a popular engagement to recreate on the tabletop. The parity of the forces involved (and the high casualty rate) allows for players to fight a very conventional meeting engagement. Napoleonic wargamers see the battle in similar terms to those fought in Europe and tend to rate it as only a small brigade level action, which it was. Yet for North America, it was a large engagement.

No special rules are required to recreate this battle. Given the relatively small size of the battle (in European terms), many Napoleonic gamers can field the required units without too much difficulty. Another attraction to this scenario is that both sides can command Native American troops, which provides an added visual appeal. When gaming in smaller scales, 15mm or smaller, blue coated Portuguese troops, or even French troops, can stand in for Americans. For these reasons, it should not be surprising that this scenario has often appeared in popular wargame magazines like *Wargames Illustrated*.

Victory Conditions

Victory conditions for this inconclusive battle are difficult to establish, as both sides acquitted themselves admirably. The American forces were well-trained, and their performance on the battlefield was solid.

80. 100th Infantry Grenadier Company, author's photo.

In hindsight, the British should have not attempted to drive the Americans off of the field and instead use the defensive works to their advantage. The Americans could have benefited by the presence of the Second Brigade, especially if they appeared on the British flank, but Brown's overall conduct of the battle was far superior to that of his predecessors. Given what happened on the battlefield, both the American and British line should be treated as at the same morale level. The militia on both sides did what was expected of them, as did the Native Americans.

To gain a dramatic victory, the British need to destroy the American force, but if they are merely able to drive them from the field, they still achieve a major victory. For the Americans, a dramatic victory would require them to destroy the British force and a major victory would require them to drive them from the field. If both sides remain on the field, the battle is considered to be a draw, with the advantage going to whoever has the most troops on the field. The scenario plays for ten turns.

Further Reading

Barbuto, Richard V. *Staff Ride Handbook for the 1812 Campaigns, 1812-1814.* (Old Fort Niagara Association. Youngstown, NY. 2016)

Elting, John R., *Amateurs, To Arms!: A Military History of the War of 1812.* (Da Capo Press. New York. 1995)

Graves, Donald, *Red Coats and Grey Jackets: The Battle of Chippawa, 5 July 1814.* (J. Kirk Howard. Louisville, Quebec. 1994)

81. US Light Dragoons, courtesy of Knuckleduster Miniatures.

Lundy's Lane - 25 July 1814

Overview
Major Battle - 6,000 participants

Lundy's Lane is known for being one of the largest battles of the War of 1812 and is commonly recreated on a tabletop. The five-hour battle produced 2,000 casualties and exhausted the American advance to Ft. George.

82. Jacob Brown by John Wesley Jarvis circa 1815, National Portrait Gallery.

Introduction

Following the Battle of Chippawa (5 July 1814), the Americans followed Maj. Gen. Brown's plan to outflank the British defensive works and cross the Chippawa River further to the west. The British realized that their position would soon be outflanked and withdrew to Ft. George. The Americans advanced onto Queenston and from there Gen. Brown observed the British defenses at Ft. George. Expecting the timely arrival of the American flotilla and heavy artillery to lay siege to the fort, Brown learned that Commodore Chauncey never left the safety of Sackett's Harbor and that he was on his own.

Lack of naval support did not stop Brown from offensive actions. He could lay siege to Ft. George and in doing so, stop the flow of supplies to Ft. Niagara, which was now in British hands. However, as the British could still supply Ft. Niagara by the lake, such an operation would have limited results. A bolder plan would be to attack and capture Burlington Heights. Though the Americans would not be able to hold the Heights, its capture would disrupt the westward flow of British supplies to places like British controlled Ft. Mackinaw. Furthermore, such an advance may draw the British out from Ft. George, where they could be destroyed. Brown returned to his camp at Chippawa and began preparing his troops to march on Burlington.

The British, learning of the American advance toward Lake Ontario, began sending reinforcements to the Niagara Peninsula. Lt. Gen. Sir Gordon Drummond, along with newly arrived regiments reinforced Ft. Niagara and began to threaten American supply lines. Though Brown was unable to quickly send troops back across the Niagara River to guard his supplies, Brown could send out a force toward Queenston. Brown ordered Scott to take his brigade forward, which Scott interpreted as a license to engaged and destroy. Drummond, who had now crossed into Canada, moved to attack the Americans.

The Battle

The Battle of Lundy's Lane, as known as the Battle of Niagara, was fought within earshot of Niagara Falls. Even in the 21st century, the sounds of the rushing water drowns out the sounds of cars, airplanes and the thousands of tourists who flock to one of the seven natural wonders of the world. In the 19th century, the sound of the falls was deafening, and its low rumble could be heard across the battlefield.

As Scott brought up his brigade, he received an intelligence report at Willson's Tavern, a local landmark that overlooked the Falls, stating that 1,100 British troops were ahead of him at Lundy's Lane. On a small rise that dominated the battlefield, the British deployed their artillery, which was well supported by infantry. Believing he could dispatch the British troops directly ahead of him, Scott deployed his brigade and quickly came to realize he was outnumbered. After a period of time, the outnumbered Americans were running low on ammunition and Scott's brigade was being devastated by long range fire. The Twenty-Fifth US Infantry had worked its way through the woods that flanked the east side of the battlefield and having dispatched a British militia unit, captured the wounded British Gen. Riall and secured the crossroads of Lundy's Lane and the Portage Road.

As the remnants of Scott's Brigade continued to hold ground, more American troops arrived on the battlefield. With daylight quickly fading, the fresh US troops formed in the rear and then advanced on the British battery. In the growing darkness, the Americans moved toward the guns. With minimal visibility, the Americans approached the guns from two sides. Though the British were able to repulse one portion of the attack, the Twenty-First US Infantry was able

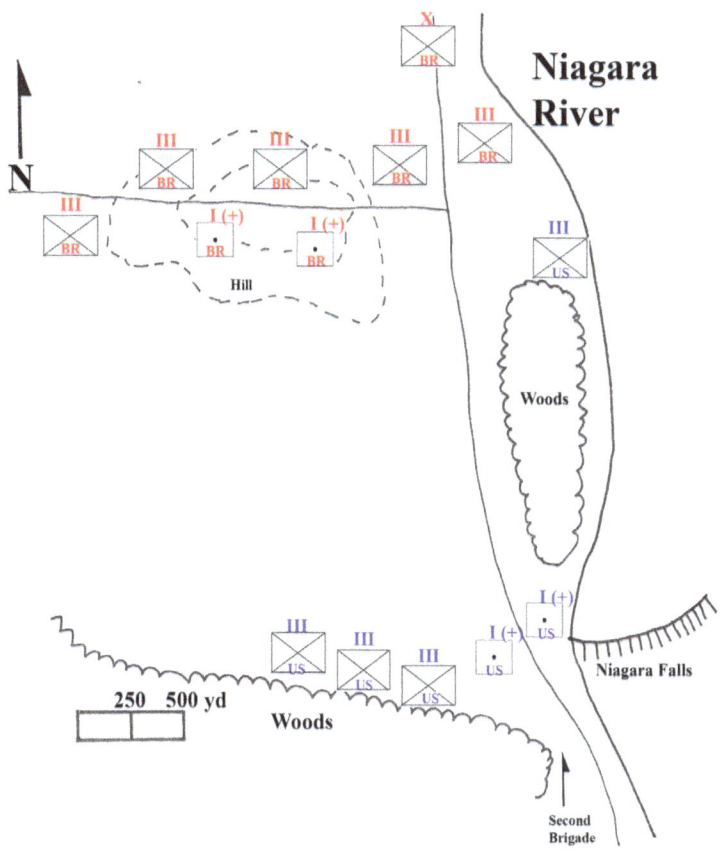

Map 35

to approach the gun unseen. As the moonlight silhouetted the gunners, the Americans let of a volley and then charged the guns. In the melee that followed, the Americans took the ridge and reinforced it with more artillery.

At this point, the British commander Drummond was wounded and British reinforcements, recently arrived on the battlefield, marched in column into the center of the fighting. Fired upon by muskets and artillery, the column broke and British began shooting at each other. Drummond, who held a belief that the Americans were undisciplined, continued to send attacks to recover his lost guns. Scott brought up the remnants of his brigade and sought to break the British line but was fired upon by American troops. At this point, Gen. Brown, who had previously been wounded, collapsed from the loss of blood.

In the darkness, the Americans withdrew their guns from the ridge and returned to their camp. The British returned in daylight and claimed ownership of the battlefield. With Brown evacuated to a hospital in Buffalo and Scott out of the war due the wounds he recently suffered the responsibility of command fell on Brig. Gen. Ripley. Ripley was ordered by Brown to withdraw the depleted US force to Ft. Erie and hold the last American bastion in Upper Canada.

Order of Battle
British
Commanding Officer: Lt. Gen. Sir Gordon Drummond

British Right Division - 3,638 men
First Brigade, 1,070 men
Lt. Col. Hercules Scott
8th Regiment of Foot, 275 men

83. Lundy's Lane Battlefield, author's photo.

84. 19th Light Dragoons, courtesy of Perry Miniatures.

103rd Regiment of Foot, 635 men
104th Regiment of Foot, 120 men
Royal Artillery, 40 men
Three 6-pdr. guns

Second or Light Brigade, 857 men
Lt. Col. Thomas Pearson
19th Light Dragoons, 95 men
Provincial Light Dragoons, 30 men
Glengarry Light Infantry Fencibles, 376 men
Volunteer Battalion of Incorporated Militia, 336 men
Royal Artillery, 20 men
    Two 6-pdr.
    One 5.5-inch howitzer

First Militia Brigade, 300 men
Lt. Col. Love Parry
Detachments from the 1st, 2nd, 4th, and 5th Lincoln Militia Regiments
Detachment from the 2nd York Militia Regiment

Second Militia Brigade, 250 men
Lt. Col. Christopher Hamilton
Detachments from the 1st and 2nd Norfolk Regiments, 1st Essex Regiment, 1st Middlesex Regiment, and Caldwell's Rangers

Lt. Col. Morrison's Force from Forts George and Niagara, 761 men
1st Regiment of Foot, 171 men
8th Regiment of Foot, 65 men
41st Regiment of Foot, 60 men
89th Regiment of Foot, 425 men
Royal Artillery, 40 men
    Two 24-pdr. guns
    Congreve Rocket Section

Division Reserve, 400 men
Lt. Col. John Gordon
1st Regiment of Foot

US
Commanding Officer: Maj. Gen. Jacob Brown

US Left Division - 2,778 men
First Brigade, 1,080 men
Brig. Gen. Winfield Scott
Ninth US Infantry Regiment, 200 men
Eleventh US Infantry Regiment, 200 men
Twenty-Second US Infantry Regiment, 300 men
Twenty-Fifth US Infantry Regiment, 380 men

Second Brigade, 882 men
Brig. Gen. Eleazer Wheelock Ripley
First US Infantry Regiment, 150 men
Twenty-First US Infantry Regiment, 432 men
Twenty-Third US Infantry Regiment, 300 men

Third Brigade, 546 men
Brig. Gen. Peter B. Porter
Regiment of New York State Detached Militia, 250 men
Fifth Pennsylvania Regiment (Fenton's Pennsylvanians), 246 men
Willcock's Canadian Volunteers, 50 men

Corps of Artillery, 200 men
Maj. Jacob Hindman
Capt. Thomas Biddle, three 12-pdr. guns
Capt. Joh Ritchie, two 6-pdr. guns and one 5.5-inch howitzer
Capt. Nathan Towson, two 6-pdr. guns and one 5.5-inch howitzer
Capt. Alexander Williams, three 18-pdr. guns
Lt. David Douglas's Company of Sappers, Bombardiers and Miners, two 18-pdr. guns
Light Dragoons, 70 men
Capt. Samuel D. Harris's Company of US Light Dragoons
Capt. Claudius Boughton's Company of New York Volunteer Dragoons

Game Notes

One can divide the battle into three phases: When Scott's Brigade engaged the British Division; when Brig. Gen. Ripley's Brigade arrived and launched their attack on the British guns on the ridge; and when the British tried to recapture their guns. In each of these separate phases were unique, both in the number of troops involved and the level of visibility. Though Scott could have possibly reduced the number of casualties his bri-

85. 11th US Infantry, courtesy of Knuckleduster Miniatures.

gade suffered at the outset of the battle by retiring into the nearby wood line, the actions of the Twenty-Fifth US Infantry during the initial phase were critical to the outcome of the final battle.

The remaining two phases of the battle were fought in the dark. Given the shared profile of the American and British uniforms, especially the newer style headgear, it was difficult to distinguish which troops were American and which were British. The arrival of new troops on the field added to the overall confusion undoubtedly contributed greatly the numerous friendly fire incidents.

Any attempt to recreate a nighttime battle on the tabletop needs to incorporate the blue-on-blue incidence of friendly fire that so commonly occurred during the last two phases of the Battle of Lundy's Lane. Given how frequently both sides suffered from friendly fire, any unit that shoots must fire see if it fired at an enemy unit or a friendly. To replicate this on the tabletop, any unit that fires needs to roll a D6 – on a roll of a 1, 2, or a 3 the unit will end up shooting the closest unit from their side. If a unit is charging, that unit must also roll a D6 and on a role of a 1 or a 2, it charges a friendly unit.

Victory Conditions

Whichever side holds the ridge at the end of twenty turns wins the game.

Further Reading

Barbuto, Richard, *Niagara 1814: America Invades Canada*. (University of Kansas Press. Kansas. 2000)
Feltoe, Richard, *A Crucible of Fire: The Battle of Lundy's Lane, July 25, 1814*. (Dundurn Press. Toronto. 2014)
Graves, Donald E., *Where Rights and Glory Lead!: The Battle of Lundy's Lane, 1814*. (Robin Brass Studio. Toronto. 1999)

86. British Foot Artillery, courtesy of Perry Miniatures.

Ft. Erie, July-September 1814

Adrian O. Mandzy

Overview
Large Battle - 6,000 participants
Large Skirmish - 840 participants
Large Battle - 7,500 participants

British lay siege to Ft. Erie. Americans launch an attack on the British siege batteries.

Introduction

The Americans had captured the semi fortified Ft. Erie at the start of the 1814 Niagara campaign and began improving its defenses. The original fort could hold a small battalion, but beginning in late July, the Americans laid out a defensive perimeter able to accommodate a division.

Following the Battle of Lundy's Lane, the wounded US Maj. Gen. Brown ordered his subordinate Brig. Gen. Eleazar Ripley to hold Ft. Erie. The British commander, Lt. Gen. Drummond still recovering from the losses suffered at Lundy's Lane, did not immediately advance on the Americans. The British pause gave the Americans the time they needed to transform Ft. Erie into a substantial fieldwork. Artillery batteries were constructed and a six- to seven- foot-high breastwork, along with a wide ditch surrounding the camp, was built. Trees were cut down, sharpened, and arranged as an abatis, which were then in turn intertwined with thorns.

The earliest British reconnaissance of Ft. Erie occurred on 3 August. Recognizing the strength of the enlarged fortification, Drummond ordered British troops to land on the American side of the Niagara River and destroy the US supply bases at Black Rock and Buffalo. The presence of a battalion of the First US Rifle Regiment at Buffalo, however, halted the British raid at Conjocta Creek and forced Drummond to consider a different avenue of attack.

The Siege

While the US Navy Commander Isaac Chauncey briefly challenged the British control of Lake Ontario, Drummond had already received the needed supplies and men to begin an attack on Ft. Erie. On 13 August, the British opened an artillery bombardment of the fort and on 15 August, Drummond launched a three-pronged night attack on the fortifications. The British troops did not have the element of surprise on their side and were met with murderous fire. The British attack on the southern defenses failed to cross the man-made obstacles while the northern attack was stopped by a combination of artillery and musket fire. Only the central column was able to cross the ditch and scale the walls of the fort. Though the British had already breached the American northern bastion and put to flight, the American artillerymen holding this section of the line, they were unable to proceed further into the fort. Accurate American artillery fire prevented more British troops being sent up to reinforce the British troops holding the northern bastion. The accidental detonation of the gunpowder reserves in the northern bastion caused horrific British casualties and shocked the troops on both sides. At this point, Drummond

87. Gen. Sir Gordon Drummond by George Theodore Berthon 1883, Government of Ontario Art Collection.

called off his attack and the British retreated to their camp.

Even after this failed attack, the British were still determined to re-capture Ft. Erie. Daily the British continued to fire their artillery at the fort, at times launching more than 500 rounds in a 24-hour span. In the no man's land between the British and American lines, patrols continued to skirmish with each other and brought in prisoners and deserters. Each side labored to improve their camps and by 28 August, the British had placed a new four-gun battery 750 yards from the American line. A few days later, the British began work on a new battery, this one only 400 yards from the fort. Running low on artillery shells and aware of more New York State Militia arriving at the fort, Drummond ordered the British batteries to conserve their ammunition. Food was also running low in the British camp and the lack of tents and stoves made life difficult for soldiers living in the rain and mud. With the autumn rains continuing to soak the British, Drummond was contemplating lifting the siege.

With the arrival of 2,200 militiamen, Brown wanted to take the fight to the British. Brown knew that the militia would not be able to stand up to the British veterans who fought against Napoleon's army in a sustained firefight and took their limitations into consideration when he formulated his plan of attack. Though the Americans never viewed the British camp directly, Brown carefully interrogated each deserter to gain as much intelligence on the enemy as possible. Knowing that the British kept only one third of their army on duty at any time, Brown proposed that his men would quickly overrun the British batteries and destroy as many guns as possible before returning to the safety of the American defenses.

Brown developed his plan in secret. Learning that US Maj. Gen. Izard and 3,000 well trained regulars would soon be arriving in the Niagara Theater, Brown believed that he could destroy Drummond by trapping him between the two American divisions. Given the complexity of coordinating such a pincer movement proved to be too difficult, the combined American force could, at the very least, force Drummond out of the Niagara Peninsula.

Brown's plan required a swift attack. Dividing his assault force into two groups, Porter's command was tasked with moving along a pre-cut path through the woods and attack British batteries #2 and #3. Porter then organized his troops into an advanced force and two assault columns. The second command under Brig. Gen. James Miller was ordered to quietly move into a ravine between the American held fort and the British batteries.

At 1430, the Americans emerged from the woods and quickly overran Battery #3. Setting off the British gunpowder reserves, the two groups of Americans charged and overcame the defenders of Battery #2. In a matter of minutes, the Americans were in control of a web of connecting British trenches. Noise from the detonation of the gunpowder brought British troops from their camp and into direct contact with the enemy. In the fog and drizzling rain, the flintlock muskets failed to spark and the New York State Militia were forced to rely upon the bayonet. Scattered into small units, the Americans continued toward Battery #1.

With 2,000 fresh British troops entering the fray, the American assault began to break down. Gen. Davis was killed while leading his men, as was Col. Gibson and Lt. Col. Wood; Gen. Porter was wounded. As the Americans began to withdraw to their lines, Gen. Ripley, sent to help extradite the raiding force, was shot and thought killed. Reaching the safety of the American lines, the British ended their pursuit.

Shortly after the sortie, the British siege was reinforced by fresh troops. On 21 September, Drummond broke camp and lifted the siege. At the time, most Americans believed that it was the sortie that had broken the siege.

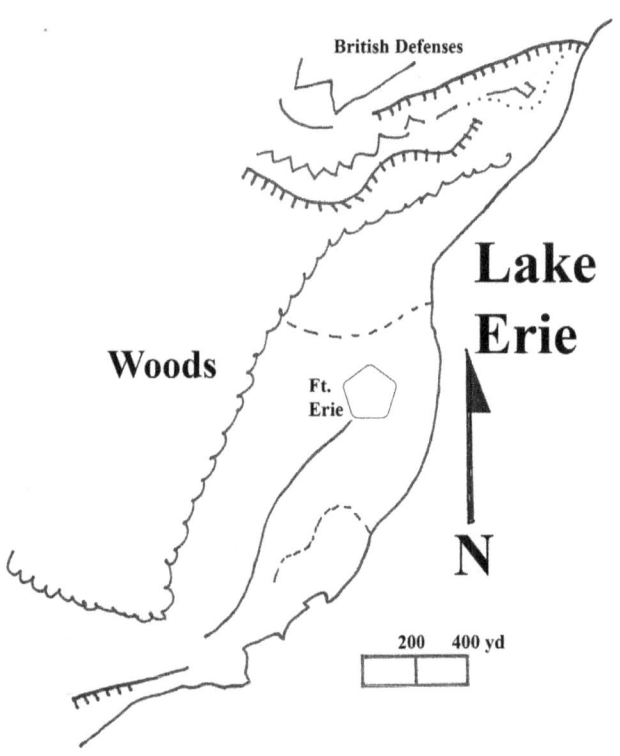

Map 36

88. Ft Erie, author's photograph.

Order of Battle
During the course of the siege, the commanders, regiments, and troop strengths radically changed. The specific orders of battle for each of the three actions are presented below.

Game Notes
The American troops, especially the New York State Militia, were not the poorly trained and inadequately led men that had sought to capture Canada in 1812. Though the lack of uniforms resulted in the New York Militia wearing read sashes around their hats, the men fought well and stood up to the British regulars on a number of occasions. A quick comparison between the numbers of American and British troops captured during the 1814 Niagara campaign, along with even a brief perusal of the numbers of documented deserters, illustrates that the American militia did not behave significantly different from the British regulars.

Scenario #1 – the British Attack on Ft. Erie, 15 August
Drummond has been criticized for failing to move on the Americans after Lundy's Lane. At that time, the American defenses remained incomplete and many troops were bivouacked in the open. A number of American commanders and officers were out of commission, say nothing of the troops under their command. By 1 August, Drummond had received a fresh battalion of British troops but having failed to disrupt the flow of supplies from Black Rock and Buffalo on the 2nd, Drummond waited two more weeks before assaulting Ft. Erie.

Drummond's continued underestimation of the Americans thwarted his goal of pushing the Americans into the waters of Lake Erie and the Niagara River. Poor British reconnaissance failed to recognize and properly assess the strength of the American field defenses between Snake Hill and Lake Erie. Drummond also vacillated when he launched the attack itself. British troops were ordered to remove the flints from their muskets, a common occurrence when launching a surprise attack, but he also ordered a demonstration by Native Americans against the center of the American defenses. The sounds of Native American war cries undoubtedly rallied every American to arms, giving them a distinct advantage when firing at the unarmed British assault teams. Drummond also failed his troops during the attack itself. When British reinforcements were needed at the northern bastion, none came up.

Conversely, the American commander during the siege did his job well. Reinforcements were sent in at the right time and the British who managed to enter the northern bastion were contained. One could argue that Ripley should have ordered a pursuit of the retreating British, but one must take into account the effects of the detonation upon the soldiers. The shock of the explosion not only brought about a lull in the attack, but effectively ended the assault. The American troops, relieved to be alive, were in no condition to go after the British as they withdrew to their camp.

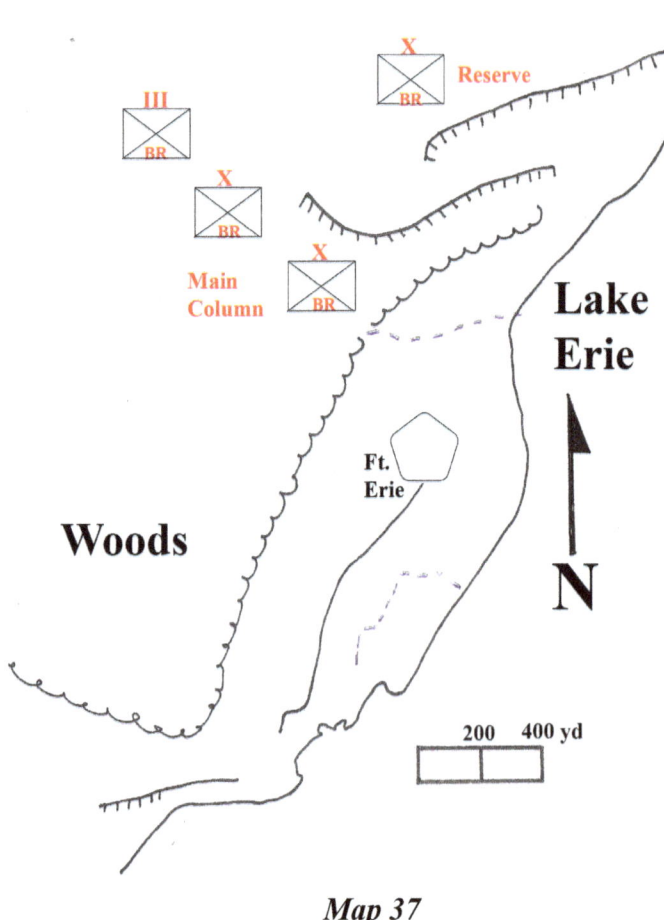

Map 37

Order of Battle
British - 3,460 men
Commanding Officer: Lt. Gen. Sir Gordon Drummond

Main Attack Column, 1,840 men
Lt. Col. Viktor Fischer
De Watteville's Regiment, 1,000 men
8th Regiment of Foot, 670 men
Flank Companies, 89th Regiment of Foot, 90 men
Flank Companies, 100th Regiment of Foot, 80 men

Supporting Attack Column against Fort Erie, 340 men
Lt. Col. William Drummond
Flank Companies, 41st Regiment of Foot, 110 men
104th Regiment of Foot, one company, 77 men
Royal Navy Sailors, detachment, 90 men
Royal Marines, detachment, 50 men
Royal Artillery, detachment, 13 men

Supporting Attack Column against the Douglass Battery, 750 men
Col. Hercules Scott
103rd Regiment of Foot, 750 men

Reserve, 1,577 men
Lt. Col. John Tucker
1st Regiment of Foot, 535 men
Glengarry Light Infantry Fencibles, 492 men
Volunteer Battalion of Incorporated Militia, 350 men
19th Light Dragoons, one company, 100 men
Royal Artillery, detachment, 100 men

US - 2,600 men
Commanding Officer: Brig. Gen. Edmund P. Gaines

Left, 620 men
Brig. Gen. Eleazar Ripley
Twenty-First US Infantry Regiment, 250 men
Twenty-Third US Infantry Regiment and an attached company of the Seventeenth US Infantry Regiment, 250 men
Capt. Nathan Towson, 120 men
Six 6-pdr. guns

Center, 955 men
Brig. Gen. Peter B. Porter
Fifth Pennsylvania Volunteers, 75 men
Swift's Regiment New York Militia, 430 men
Combined First and Fourth Rifles, 350 men
Capt. Alexander C. W. Fanning, 40 men
    Two 6-pdr. guns
Capt. Thomas Biddle, 60 men
    Three 6-pdr. guns

Fort Erie Garrison, 198 men
Maj. William A. Trimble
Nineteenth US Infantry Regiment, 118 men
Capt. Alexander J. Williams (80 men)
    One 24-pdr. gun
    One 18-pdr. gun
    One 12-pdr. gun

Right, 455 men
Lt. Col. William McRee
Ninth US Infantry Regiment, 185 men
Combined Pennsylvania and New York Volunteers, 120 men
New York Dragoons, 100 men
Bombardiers, Sappers, and Miners, 50 men
    One 6-pdr. gun
    One 12-pdr. gun

Reserve, 366 men
Lt. Col. Thomas Aspinwall
Eleventh US Infantry Regiment, 130 men
Twenty-Second US Infantry Regiment, 160 men
Company US Light Dragoons, 56 men

89. New York State Militia wearing their distinctive red headscarves, author's photo.

Scenario Victory Conditions
    The British need to capture the fort and destroy or drive out the Americans. This scenario plays for twelve turns.

Scenario #2 – Skirmish of 12 August
    During the course of the siege, numerous skirmishes were conducted in the no man's land between the American and British camps. Such *petite guerre* encounters were a common everyday occurrence and were launched by both sides to gather prisoners, gain intelligence, keep up morale among the troops, and wear down the enemy. At Ft. Erie, the Americans launched raids in order to delay the British construction of their siege batteries.
    The skirmish of 12 August was typical of such an engagement. Though in this particular skirmish Ludo-

wick Morgan, the commander of the First US Rifles, lost his life, such actions typified the continued loss of life on both sides.

Game Notes

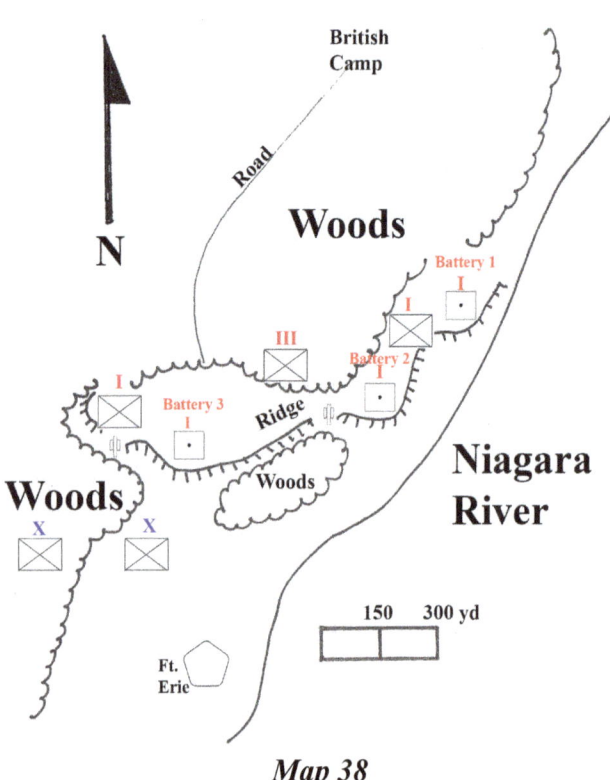

Map 38

The goal of the American patrol is to disrupt the British construction of their battery. The longer the Americans maintain a presence in the field, the less work the British are able to accomplish. If the Americans are within 12 inches of the uncompleted battery, they automatically gain a victory point. At that point, the British work parties will automatically withdraw six inches to the rear, even if they are not being shot at by the Americans. The longer the Americans threaten the work party, the more points the Americans gain. If the Americans are able to attack the work parties and cause any casualties, they gain an additional victory point for each work party member killed or captured.

However, the longer the Americans remain on point, the more British troops arrive to deal with the American threat. For every turn the Americans are within 12 inches of the battery, the British player rolls a D6. On a roll of a 5 or a 6, an additional British company will arrive. A second die roll determines which British company arrives at the edge of the table. The British gain a victory point for every two turns the Americans are not within 12 inches of the uncompleted battery. In addition, the loss of a commanding officer on either side gives the enemy an additional victory point.

Order of Battle
British
Commanding Officer: none

In the field
Glengarry Fencibles, two companies, *120 men*
Volunteer Battalion of Incorporated Militia, one company, *60 men*
John Norton, leading a group of Iroquois Mohawks, one warband, *40 men*

At the uncompleted battery
Lt. George Phillpotts, Officer of Engineers
Work party, 20 men

Reinforcements
On a roll of a 1 on a D6 – DeWatteville Regiment, one company, *60 men*
On a roll of a 2 on a D6 - Volunteer Battalion of Incorporated Militia, one additional company, *60 men*
On a roll of a 3 on a D6 – 1st Regiment of Foot, one company, *60 men*
On a roll of a 4 on a D6 – 104th Regiment of Foot, one flank company, *60 men*

90. US Infantry, author's photo.

On a roll of a 5 on a D6 – Royal Marine, one company, *60 men*
On a roll of a 6 on a D6 – 19th Light Dragoons, one troop, *60 men*

US
Commanding Officer: Maj. Ludowick Morgan
First US Rifles, two companies, *120 men*
Fourth US Rifles, two companies, *120 men*
New York State Militia, one company, *60 men*
Canadian Volunteers, one company, *60 men*

Scenario Victory Conditions
   The side with the most victory points wins the game. The game plays for eight turns.

Scenario #3 - The Sortie
Intro
   Despite the poor weather, Brown actively wanted to break the siege and take the initiative. By destroying the British guns, the American position at Ft. Erie would no longer be seriously threatened and the US forces could regain the initiative.

Game Notes
   This in an evenly matched encounter, with the Americans having the slight numerical advantage at the start. With the destruction of the first British battery, the British need to get their men out of camp and defend their guns. The Americans need to quickly reach the remaining batteries and destroy the cannon before the entire British force arrives. The game begins with Porter's force deployed within charge distance of the Battery #3. Millers troops will only become engaged after Battery #3 is destroyed. Ripley's men become active only after battery #2 is destroyed.
   To replicate the Americans getting lost in the mist and the iumble of trenches between the batteries. each time an American unit moves forward, it needs to roll a D6. On a roll of a 1, the unit rolls a second D6 and on a roll of a 1,2, or 3, it moves it full movement to the right and on a roll of a 4, 5, or 6, it moves its full movement to the left of its current position.
   Most of the British units are in camp, with only the Duty Brigade deployed among the three batteries. Additional British units can begin to move forward only when the Americans blow up Battery #3. The first British group can move from the camp at the start of the turn following the destruction of the first battery, two turns later the second British group can move from the camp, three turns after the detonation the 3rd group become active. The British Reserve can enter the fray at the start of turn nine.
   The game begins with the Americans within charge distance of Battery #3. It takes a turn for a unit to destroy a battery.

Order of Battle
British - 4,040 men

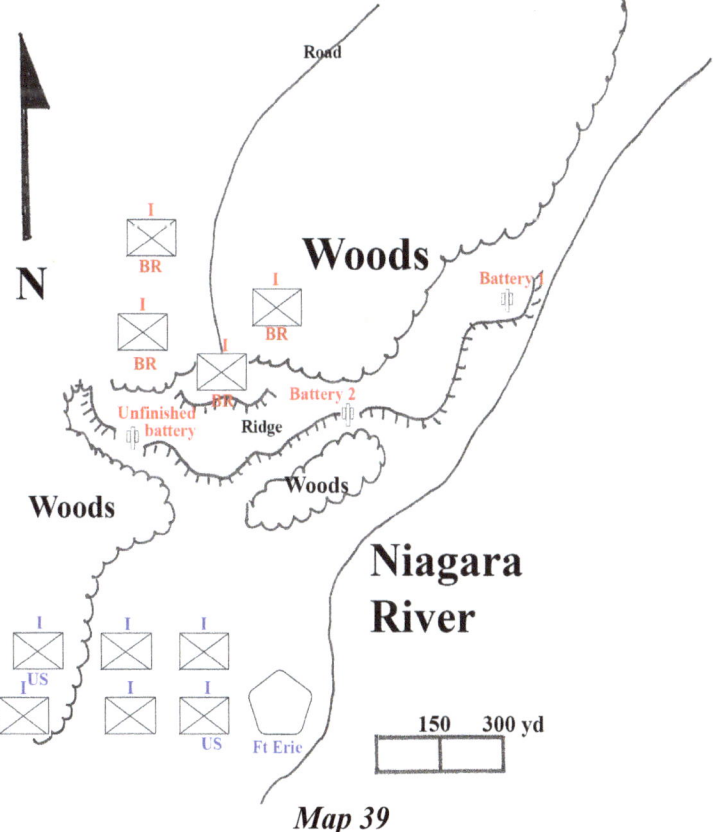

Map 39

Commanding Officer: Lt. Gen. Sir Gordon Drummond

The Duty Brigade, 1,200 men
Lt. Col. Viktor Fischer
De Watteville's Regiment, 600 men
8th Regiment of Foot, 500 men
Royal Artillery, 100 men
    Three 24-pdr. guns
    One 18-pdr. gun
    One 64-pdr. carronade
    One 10.5-inch mortar

The British counterattacked with three groups.
Lt. Col. John Gordon's Group, 700 men
1st Regiment of Foot
89th Regiment of Foot

Maj. Gen. Louis De Watteville's Group, 1040 men
6th Regiment of Foot, three companies
82nd Regiment of Foot

91. 1st Royal Scotts, courtesy of Knuckleduster.

Lt. Col. Francis Battersby's Group, 300 men
Glengarry Light Infantry Fencibles

British Reserve
Lt. Col. Campbell, 900 men
6th Regiment of Foot, seven companies
Flank Companies 41st Regiment of Foot
Volunteer Battalion of Incorporated Militia
19th Light Dragoons

US - 3,370 men
Commanding Officer: Maj. Gen. Jacob Brown

The Assault Force, 1,670 men
Brig. Gen. Peter B. Porter

Advance Guard, 200 men
Col. James Gibson
Combined First and Fourth Rifles, 180 men
Seneca Warriors, 20 men

Right Column, 970 men
Lt. Col. Eleazar Wood
Combined First and Twenty-Third US Infantry Regiments, 400 men
US Light Dragoons (dismounted), one company, 70 men

New York State Militia, 500 men

Left Column, 500 men
Brig. Gen. Daniel Davis
New York State Militia, 500 men

Miller's Detachment, 610 men
Brig. Gen. James Miller
Combined Ninth and Nineteenth US Infantry Regiments, 300 men
Eleventh US Infantry Regiment, 310 men

Reserve, 640 men
Brig. Gen. Eleazar Ripley
Twenty-First US Infantry Regiment, 540 men
Seventeenth US Infantry Regiment, 100 men

Fort Erie Garrison, 450 men
Maj. Thomas Jesup
Twenty-Fifth US Infantry Regiment, 150 men
Artillerymen, 300 men

Scenario Victory Conditions

Victory conditions for the sortie depends on the number of batteries overrun and destroyed. If the British save two batteries it's a minor victory, if the Americans destroy two it's an American minor victory. If a side defends or captures all the batteries, they gain a stellar victory.

The game plays for twelve turns.

Further Reading

Barbuto, Richard
2000 *Niagara 1814: America Invades Canada.* University of Kansas Press. Kansas.

Elting, John R.
1995 *Amateurs, To Arms!: A Military History of the War of 1812.* Da Capo Press. New York.

Owen, David A.
1986 *Fort Erie (1764-1823): An Historical Guide.* Niagara Falls Parks Commission. Niagara Falls, Ontario.

Cook's Mill, 19 October 1814

Bad Roads and Poor Rations

Overview
Large Skirmish - about 1,500 engaged

In an effort to draw out British troops from their fortified positions, the Americans launch an attack on a British supply depot. It was the last American engagement fought on Canadian soil during the War of 1812.

Introduction

The first two years of the War of 1812 had not gone according to the US plan and American negotiators at Ghent learned that the British proposed that each side should retain any territory it had captured. To that end, the British began planning attacks on New Orleans, New York and Plattsburgh. The largest combat ready American group, recently reinforced by troops from Sackett's Harbor, stood at the ready on the Niagara frontier. With two trained divisions, the American army stood poised to press forward.

At this critical moment, the army learned that the American Navy had withdrawn to Sackett's Harbor. Without naval support to stop the British from landing fresh troops in the American rear, the American Army commander Maj. Gen. George Izard believed that any additional attacks would be fruitless. Izard then brought his army toward the British defensive line at Chippawa and after engaging in an artillery duel, withdrew behind Black Creek.

Feeding the large British army at Chippawa was difficult. If the food supply was further reduced, Izard reasoned, the British forces may withdraw to be closer to the Kingston supply line. Before going into winter quarters, Izard ordered Bissell's Brigade to either capture or destroy the flour at the mill.

Moving over bad roads and destroyed bridges, Bissell's force arrived at the mill and quickly drove off the twenty men stationed there. Fearful of being outflanked, the overall British commander Gen. Sir Gordon Drummond dispatched a force under Col. Christopher Myers to assess this new threat.

The Battle

On the morning of the 19th, the British commander Col. Myers formed his troops a mile away from the Americans. Placing his Glengarry light troops in the front line with the remaining line troops moving in support, the British drove the American Riflemen back toward the mill. Responding to the British advance, the Americans formed their line units in support and drove the Glengarries back. The Americans pushed the British through a small wood. The British fell back through a small ravine and set up a line on its crest. Upon reaching the edge of the woods, the Americans continued to fire at long range. The British brought up their cannon and fired rockets at the Americans, but the US troops did not advance from the cover of the trees. Seeing that the Americans were unwilling to advance further, the British withdrew from the field. Bissell, believing he had driven off the British, proceeded to destroy 200 bushels of grain and withdrew to the American lines.

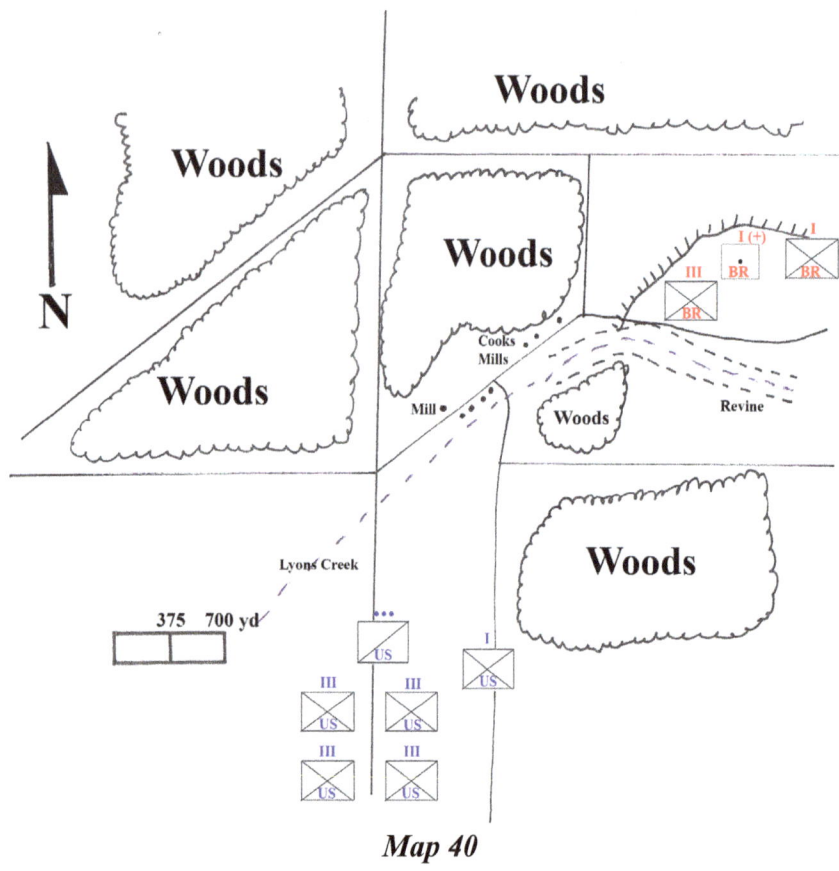

Map 40

- 130 -

Order of Battle

British - 750 men
Commanding Officer: Col. Christopher Myers

82nd Regiment of Foot, *300 men*
100th Regiment of Foot, *250 men*
Flank Companies, 104th Regiment of Foot, *120 men*
Glengarry Light Infantry Fencibles, *60 men*
Royal Artillery, One 6-pdr. gun, *20 men*
Rocket Corps, Detachment, *20 men*

US - 900 men
Commanding Officer: Brig. Gen. Daniel Bissell

Fifth US Infantry Regiment, *200 men*
Fourteenth US Infantry Regiment, *200 men*
Fifteenth US Infantry Regiment, *200 men*
Sixteenth US Infantry Regiment, *200 men*
Fourth Rifle Regiment, one company, *60 men*
US Light Dragoons, detachment, *24 mounted men*

92. Daniel Bissell, courtesy John R. Lee

Game Notes

The troops on both sides are competent and the commanders on both sides failed to make any mistakes. The British recognized that defending the crest of a ravine would be advantageous, as did the Americans. As no one wants to be last to die in a war that is coming to an end, the men and officers of both sides did their duty but did not push to cover themselves in glory.

The scenario is easy to recreate on the tabletop. No special game mechanisms are required, and the small number of troops engaged makes it easy to field at a scale of 1 to 10 or even 1 to 5. The American Dragoons were not engaged while the British cannon and Congreve Rockets offset the slight American numerical advantage, especially when defending a ridge line or the crest of a ravine.

Victory Conditions

For the British or the Americans to obtain a glorious victory, they need to drive the enemy from the field by the end of the game. Any other outcome is considered to be a draw. The game plays for twelve turns.

Further Reading

Barbuto, Richard, *Niagara 1814: America Invades Canada*. (University of Kansas Press. Kansas. 2000)

Collins, Gilbert, *Guidebook to the Historic Sites of the War of 1812*. (Dundurn Press. Toronto. 1998)

Fredricksen, John D., *Green Coats and Glory: The United States Regiment of Riflemen, 1808-1821*. (Old Fort Niagara Association. Youngstown, New York. 2000)

93. Cook's Mill Monument, courtesy John R. Lee

94. Glengarry Light Infantry, courtesy of Tod Kershner.

95. US Rifle Regiment, courtesy of Roger Chrysler.

Western Great Lakes

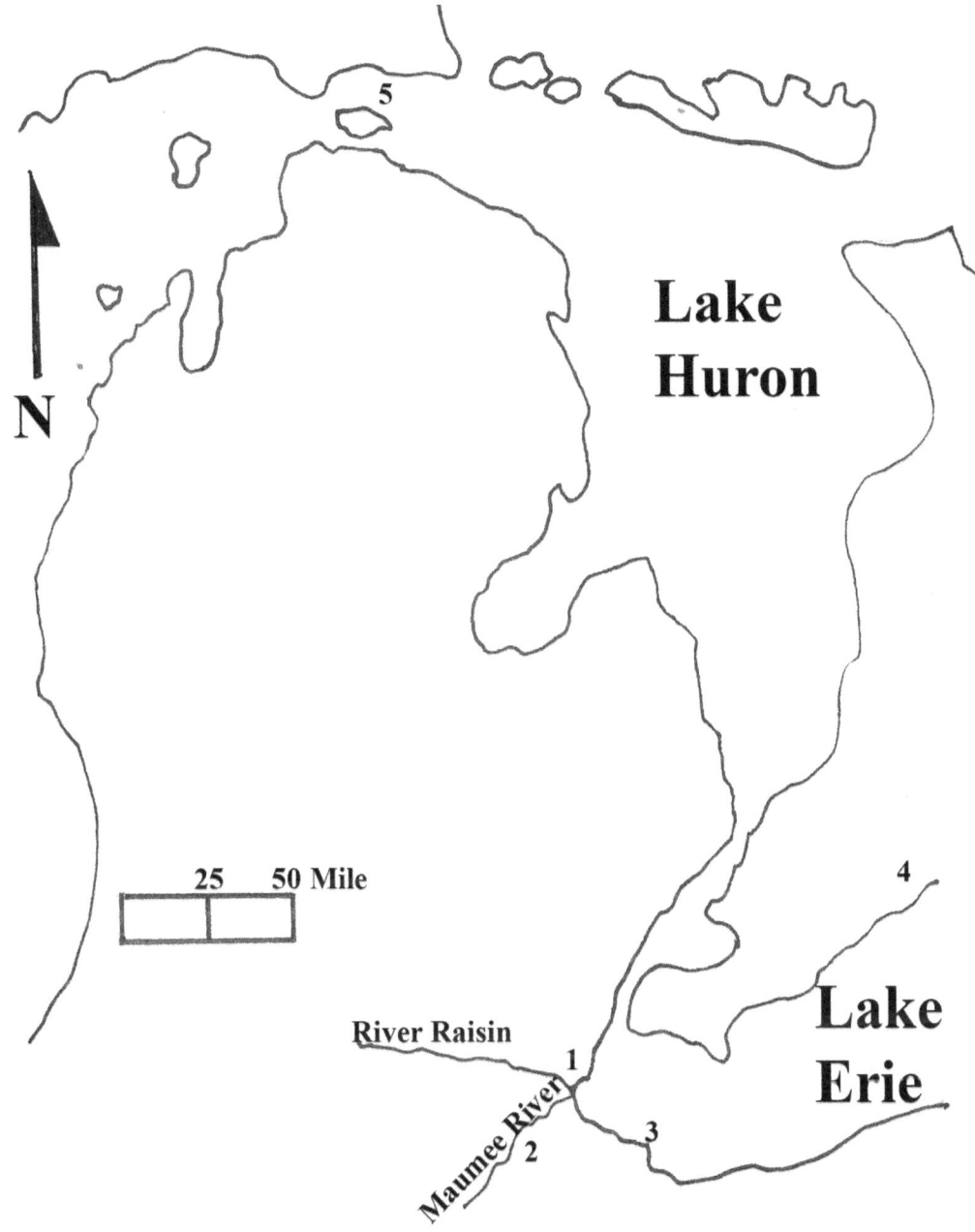

1. River Raisin   2. Ft. Meigs   3. Put-In-Bay   4. Thames   5. MacKinac Island

Map 41

The River Raisin, 22 January 1813

Overview
Battle - approximately 2,000 participants

British and Native Americans attacked and captured the town of Frenchtown.

Legacy
Battle of the River Raisin was best remembered not for the battle itself, but for the Native American massacre of American wounded prisoners in the days that followed. It has been estimated that sixty wounded men were killed by the Native Americans - some outright while others were first subject to torture. Stories of the recent Native American behavior at the River Raisin played into the preexisting fears and anxieties of the white settlers, which further hardened their stance against the Native Americans. During the course of the war, "Remember the River Raisin" became a battle cry for Kentucky troops and was used to reinforce the idea of not giving mercy to any Native Americans.

Introduction
The start of the war was not going well for the Americans, as poor American communications gave the British a distinct advantage in the opening months of the conflict. Gen. Hull's army was still on route to Detroit when he was told to march with all possible speed to his destination. Hull was able to speed up his progress by chartering the schooner *Cuyahoga* to ferry his baggage and headquarter records. Unfortunately for Hull, when the *Cuyahoga* fell into enemy hands, the British gained an intelligence goldmine. To add insult to injury, the British were able to seize the northern most American outpost of Mackinac without the Americans even firing a shot in their defense.

96. Frenchtown Battlefield circa 1860 (Lossing).

Native Americans now threatened American settlements throughout the territory and Detroit was beginning to be overrun by refugees. Detroit itself was low on food and supplies, with its lines of supply threatened. For the Americans, what was supposed to be just a matter of marching into Canada was turning into a comical farce, or a nightmare.

Hull did cross into Canada but faced with mutinous American militia and his superior Gen. Dearborn entering a truce with his British counterpart, Hull's invasion accomplished little. Hull's British counterpart, Gen. Isaac Brock learning that Hull's army was weak, poorly supplied and full of internal dissension, decided to press on Hull. Brock set up a battery across the river in the town of Sandwich and called on Hull to surrender. Having witnessed Native American attacks on civilians and prisoners during the American War of Independence, Hull initially resisted the British forces. Brock was able to partially encircled Detroit and following a two-day bombardment by both the British land battery and by ships in the Detroit River, Hull surrendered.

Hull's surrender shifted the initiative to the Native Americans. Fearful of the complete collapse of frontier settlements, the governors of Ohio and Kentucky called up new recruits. James Winchester, a Revolutionary War veteran, was appointed general and marched his Kentucky troops north. In November 1812, the British had established an advanced outpost at Frenchtown, a small settlement south of Detroit. The Americans citizens of Frenchtown reached out to the Americans and asked for their help in driving out the British and Native American forces from Frenchtown. In January 1813, American troops under Col. John Allen and Col. William Lewis crossed the frozen River Raisin and pushed the British militia and Native Americans out of the settlement. Gen. Winchester then decided to hold the town and ignored warnings of an imminent British counterattack.

The Battle

The British commander, Col. Henry Procter, gather a combined force of regulars and Native Americans and began to move on Frenchtown. Resting for the night five miles from Frenchtown, Procter hoped to assault the settlement at dawn. The town itself was too small to hold the American force and the Seventeenth US Infantry Regiment were camped outside the town.

In the early morning light, an alert American sentry spotted the British as they were forming up to attack the palisaded settlement. Awoken to the sounds of gunfire, the American militia manned the puncheon fencing palisades and began to exchange fire with the British. The fire from the Kentucky troops was accurate and they were able to silence the British gun crews shooting at the palisades.

The US regulars were caught in the open and were fired upon from three sides. Charged by Native Americans, the regulars broke and began to run. Winchester was captured and turned over to Procter. Though some 400 Kentuckians still held the town, Winchester advised them to surrender, which they did. Some thirty Americans escaped from Frenchtown and reported the situation to Maj. Gen. William Henry Harrison, who was nearby. Harrison then set fire to his supplies and retreated.

Procter believing his duty done, marched his regulars back to Canada, taking the walking American prisoners with him. The Native Americans then set fire to the hospital and killed the wounded American prisoners or took them away as captives.

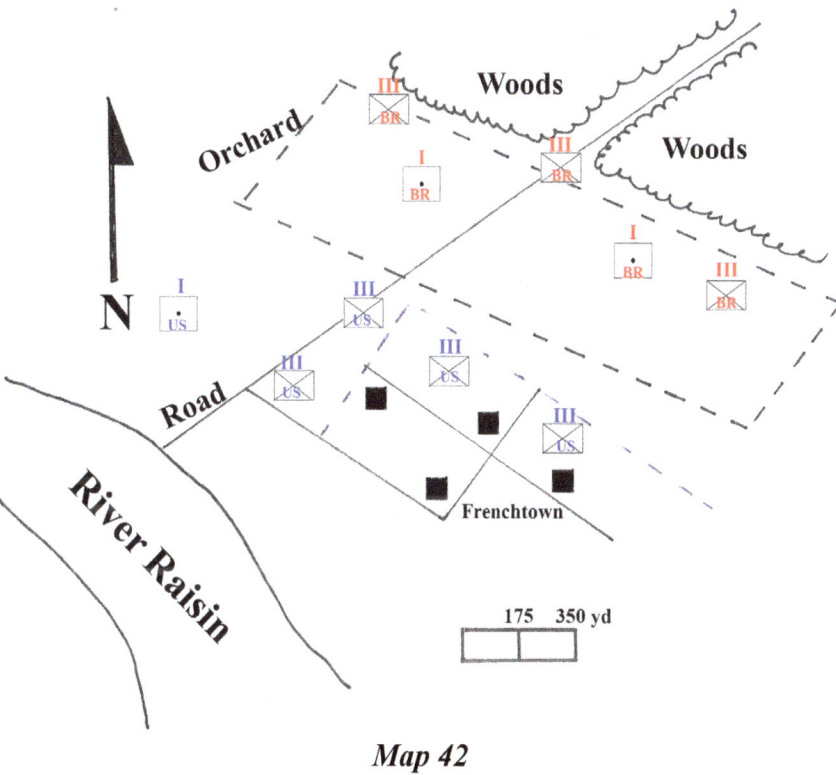

Map 42

Order of Battle
British - 1,200 men
Commanding Officer: Col. Henry Procter

Col. Procter, 600 men
41st Foot, 244 men
Royal Newfoundland Fencibles, 61 men
Royal Veterans, 4 men
1st Essex Militia, 116 men
2nd Essex Militia, 96 men
Three 3-pdr. cannon drawn on sledges
Three 3-pdr. howitzers drawn on sledges
    Royal Artillery, 21
    Canadian Provincial Marines, 28 men

Bad Roads and Poor Rations

Native warriors, 600 men
Roundhead, Wyandot war chief
Walk-In-The-Water, Wyandot war chief
Potawatomi, Shawnee, Ottawa, Chippewa, Miami, Winnebago, Creek, Mingo Iroquois, Fox, and Sauk warriors.

US - 950 men
Commanding Officer: Brig. Gen. James Winchester

Right Wing – in the open
Maj. Elijah McClenahan
Seventeenth US Infantry Regiment, three companies, 225 men
Nineteenth US Infantry Regiment, one company, 75 men

Left Wing – in Frenchtown
Lt. Col. William Lewis
1st Battalion, First Kentucky Rifle Regiment, *53* men
2nd Battalion, First Kentucky Rifle Regiment, composite unit, 54 men armed with muskets
2nd Battalion, First Kentucky Volunteer Militia Regiment, *196* men
2nd Kentucky Volunteer Militia Regiment, one company, 21 men
2nd Battalion, Fifth Kentucky Volunteer Militia Regiment, *325* men

97. Sled gun, courtesy of Ralph Naveaux.

Game Notes

The battle has two distinct phases – the initial British assault on Frenchtown and the longer running battle between the Kentucky troops protected by the settlement's palisade and the British troops. Historically, Gen. Winchester was captured and recommended that the still fighting US troops surrender, but such an action is difficult to replicate on the tabletop. Indeed, the senior commanders on both sides did little to distinguish themselves on the field of battle and both Prevost and Harrison withdrew from the point of contact.

The forces loyal to the Crown have a number of advantages. The British have a slight numerical advantage over their adversary, while the historical bivouac of the American line troops outside the town placed these men at a distinct disadvantage. Numerous accounts also state that the Kentucky troops were running low on ammunition; it is unclear if this is the result of them being in a long and sustained firefight or was this the result of Harrison or Winchester failing to bring the proper supplies up to the field after these men had captured the town. In our game, the Kentucky troops were supplied at the start of the battle but could find themselves with limited ammunition stocks following a particularly bad die roll (they rolled a one). In such a case, five chits are placed behind the unit with low ammunition. For the rest of the game, every time this unit fires, it spends a chit. Once the unit has no more chits, they can no longer fire. Chits can be transferred between friendly units.

The Americans also have a few advantages. The conduct of an attack requires solid leadership, which by all accounts Procter was lacking in. The thick snow which covered the landscape made movement difficult, which reduced the speed at which the attackers could move. The Kentucky Riflemen were also crack shots, putting out of action a number of the British guns. Not only did the riflemen have a longer reach than the standard musket, but the men behind them could place their bullets on targets. The British had suffered significant losses in attacking the Americans behind the palisades and had Winchester been killed, it is unclear if the British would have been able to storm the town.

The game begins with the British setting up to assault the American lines. The British are within two charge moves of the American held palisade. One American unit is a guard while the remainder of the American units are all disordered at the start of the game. Movement through the snow is reduced for all troops besides the Native Americans. Once the British army suffers 50% casualties, they will withdraw from Frenchtown.

98. Native Americans, courtesy of Tod Kershner.

Victory Conditions
    The Americans win if they hold the town at the end of the game. The game plays for 12 turns.

Further Reading
Elting, John R., *Amateurs, to Arms! A Military History of the War of 1812*. (Da Capo Press. New York. 1995)

Hitsman, J. Mackay, *The Incredible War of 1812*. (Robin Brass Studio. Montreal. 2012)

Naveaux, Ralph, *Invaded On All Sides*. (Walsworth Publishing Company. Marceline, MO. 2008)

99. Kentucky Volunteer Rifles, Command, courtesy of Knuckleduster Miniatures.

Ft. Meigs, 28 April - 9 May 1813

Overview
Large Battle - 4,500 participants
Small Battle - 1,720 participants
Large Skirmish - 600 participants
Large Skirmish - 850 participants
Small Skirmish - 6 participants

British troops and Native Americans lay siege to a fort. Kentucky militia attempt to lift the siege.

Introduction

Following the British capture of Detroit, several thousand troops were raised in Kentucky, Ohio, and Pennsylvania to deal with the threat. William Henry Harrison, the former governor of the Indiana Territory, was placed in command of three columns marching to retake Detroit. Following the defeat of the Americans at River Raisin on 22 January 1813, Harrison retraced his steps and began building a fort on the Rapids of the Maumee River. On 2 February 1813, ground was broken, and the fort as named after Return Jonathan Meigs, Jr, the then current governor of Ohio. Designed by Capt. Charles Gratiot, an 1806 graduate of West Point, to house 2,500 men, Harrison's command was held by only a few hundred men. In addition to blocking further British advancement into Ohio, the sprawling fort was to be a staging area for renewed American advances into Canada.

For three months, the men toiled in the mud and snow, finally completing the fort's construction at the end of April 1813. The fort, built of earth and heavy logs, sat on a bluff above the Maumee River. To the east and the west of the fort, naturally deep cut ravines protected the installation. The fort's southern approach was cleared of trees, which served to deprive any attackers from cover. Additional earthworks were constructed to protect the fort.

100. William Henry Harrison by Rembrandt Peale circa 1813, National Portrait Gallery.

Made aware of the growing American presence, the British victor of the River Raisin, Brig. Gen. Henry Procter, delayed his advance on Ft. Meigs. Procter's Army had suffered serious losses at Frenchtown and the British commander waited to be reinforced by the two flank companies of the 41st Foot. The movement of cannons necessary to conduct a siege required dry weather and Procter needed to wait until the spring rains had subsided. In mid-April, Procter began his pre-emptive advance to stop the next American invasion of Canada. Harrison, learning of Procter's plan, brought an additional 300 men to the fort and ordered a brigade of Kentucky militia to follow as quickly as possible.

On 26 April, Procter landed at the mouth of the Maumee River. For the next five days, Procter moved his cannon and gunboats up to the fort and prepared to lay siege. British gun batteries were erected on both sides of the river. Tecumseh, the Native American who accompanied Procter, led his warriors across the river and prepared to isolate the fort's southern approach.

The Battle

On 1 May, the British began to bombard the fort. For four days, the British shells fell without causing much damage, as the wet weather had turned the earthen walls into a muddy sponge that absorbed the British cannonballs. The American garrison had limited supplies and was forced to use recovered British cannon balls. When the British established a fifth battery on the southern side of Maumee River, approximately 300 yards east of the American stockade, the continued defense of the fort was in doubt. All the American hope now laid in the timely arrival of reinforcements from Kentucky.

Kentucky Brig. Gen. Green Clay, cousin of US Congressman Henry Clay, was marching 1,200 militiamen to reinforce Harrison, when a courier from the fort delivered new orders. Two thirds of the Kentucky troops under Lt. Col. William Dudley were to go downriver on flatboats and destroy the British batteries set up on the northern bank of the Maumee River. The remaining Kentucky troops under Lt. Col. William Boswell would land on the south bank, punch through the Native Americans blockade and make their way to the fort. Once the siege guns on the northern side of the river were neutralized, the fort's garrison would sally forth and attack the British battery on the southern side of the river.

On the morning of the 5 May, the Kentucky troops were in position to execute their plan. The Native Americans had posted few guards and a torrential downpour helped the Americans maintain surprise. Dudley's twelve boats landed at their appointed place and after a short engagement with the Native Americans where a few militiamen were wounded, formed into three columns and advanced on the British batteries. Dudley's men captured the batteries without resistance, but the spikes necessary to disable the guns were with the troops who landed on the other side of the river. The militia attempted to dismantle the captured cannons, but lacking the proper tools, did an ineffective job.

At this point, Native Americans began to engage the Kentucky troops. Using feint attacks, the Native Americans were able to draw the Kentucky troops inward. Fighting in close quarters for the better part of an hour, the American command collapsed, and men fought in small groups. Dudley was killed and scalped in front of his men; others were made prisoner. Of the almost eight hundred men involved in the attack, only 170 men were able to fight their way back to the boats and cross the river to the fort.

The American reinforcements that landed on the south side of the river fared much better. Facing stiff resistance, Harrison sent 200 dragoons and a company of infantry from the fort to help the reinforcements fight their way into the stockade. Once Clay was within the fort, 350 men of the Nineteenth US Infantry Regiment stormed the British battery on the south side of the river. After a fierce fight, the Americans were able to destroy the guns and return to the fort with a number of British prisoners.

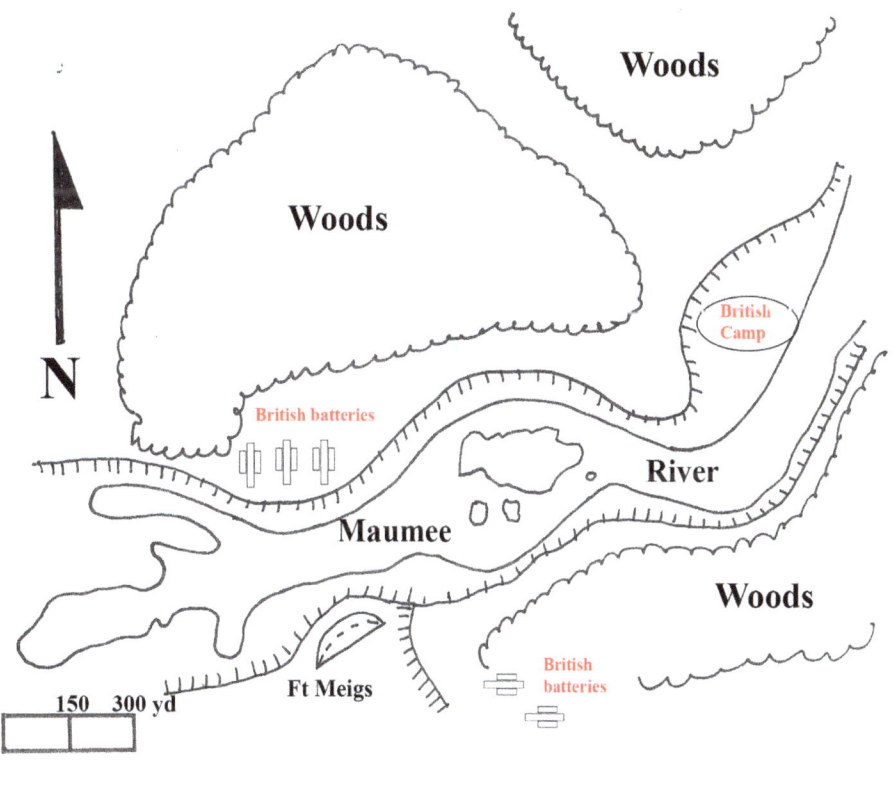

Map 43

For two days, the British and Americans discussed prisoner exchanges and on the 5th of May Procter resumed his artillery bombardment. As the Native Americans had gained trophies, loot, and prisoners, many of them drifted off from the battlefield. The British militia, concerned with the planting season, protested that they needed to return to their farms. With more rain, the initiative for maintaining the siege was drowned out. Procter embarked his troops and Harrison did not pursue.

Though the fort was again threatened two months later in July when Tecumseh tried to capture the installation by way of an elaborate ruse, the American commander was not deceived, and the fort did not fall. Lacking heavy artillery, the British were unable to storm the fort directly.

Order of Battle
British - 979 men, not including the 1,200 men under Tecumseh
Commanding Officer: Brig. Gen. Henry Procter
41st Foot, 423 men
Royal Newfoundland Regiment, 63 men
Royal Artillery, 31 men
Militia, 462 men

Battery #1
Two 24-pdr. guns
One 8-inch howitzer

Battery #2
Two 12-pdr. guns

Battery #3
Two 9-pdr. guns

Mortar Battery
Two 5.5-inch mortars

Tecumseh
1,200 men

101. Ft Meigs, author's photo.

US - *2,310 men*
Commanding Officer: Maj. Gen. William Henry Harrison

At the fort, *1,050 men*
Seventeenth US Infantry Regiment, *120 men*
Nineteenth US Infantry Regiment, 120 men
Dragoons, two companies, 200 men
Militia, *610 men*
Seven blockhouses
Four gun batteries
One mortar battery
Eighteen guns, including 18-pdrs., 12-pdrs., and 6-pdrs.

Reinforcements 1,260 men
Commanding Officer: Brig. Gen. Green Clay
Lt. Col. William Dudley

800 men
Twenty-Fourth US Infantry Regiment, one company, 60 men
Lt. Col. William Boswell
400 men

Game Notes

The highpoint of the siege is when the Kentucky troops reach Ft. Meigs. Harrison's plan on using Clay's reinforcements required timing, coordination, and military discipline – something the Kentucky troops were lacking. Dudley's men were clearly unaware that after they captured the guns, they were to return to their boats and make it to the fort. Existing accounts from the Kentucky survivors all point to the absence of discipline and a lack of even a basic understanding of Native American warfare. For the Kentucky troops raised on the stories of the Native American victory at Blue Licks, the education was wasted, as the men at Ft. Meigs fell into the same tactical trap.

102. Native Americans, courtesy of Roger Chrysler.

For gamers in the 21st century, the commonly held values of the early 19th century are significantly different. Gamers understandably focus on the "fun" part of the hobby, be it painting, rolling dice with friends, or simply picking things apart. While certain concepts such as honor and slavery are familiar to the majority of gamers, the early American fascination with scalps, be it gaining them or finding them among enemy prisoners, led to a number of poor choices. Much like among Native Americans, where the concept of counting coups was a critical component of society, militia troops, especially from Kentucky, obsessed with collecting scalps. American troops are reported to have dug Native American graves in search of scalps. Scalps were so valued that George McFeely of Kentucky, when made prisoner by the British, hid two Native American scalps on him until he was paroled. In 1814, it was reported that Kentucky troops also took scalps of British troops that fell under their knives.

For game purposes, especially in this scenario, Kentucky troops need to check if they charge Native Americans in pursuit of scalps, regardless of their training. When Kentucky militia are fired upon by Native Americans, they need to roll a five or less on a D6. The following modifiers are cumulative.

Unit has suffered casualties in pervious turns
Unit has suffered casualties in this turn
Unit is below 50%
Officer killed

For each condition that applies, the required die roll becomes more difficult. If all the modifiers are applicable, the unit would need to roll a one of the D6. If they fail their die roll, they will charge the closest Native American unit. If they fail to reach their target, they become disordered and will continue to charge next turn.

While it is possible to play the entire event on the tabletop, the three separate actions occur independently from each other, and each would make for a balanced tabletop recreation. If one were inclined to recreate the entire relief of Ft. Meigs, splitting the game onto three tables would reduce the need to model the sprawling fortification. A fourth scenario, independent of the other three events, is also included.

Scenario #1 - Dudley's Defeat

The American player needs to land their troops and destroy the four (McAfee says three) British batteries. Formed into three columns, the Americans advanced quietly. When the Americans were fired upon by a small group of Native Americans, the Kentucky troops cheered and launched a charge. The US troops captured the guns without any loss of life, but the lack of proper tools resulting in their inability to put the guns out of action.

Harrison's plan was for the Kentuckians to destroy the guns and return to the fort, but unfortunately

Dudley did not share these plans with his junior officers. Aware of the presence of Native Americans, the militia engaged their enemy. Discipline broke down and the Americans fell into a trap from which few escaped.

For our game, the scenario begins with the American troops having seized the guns. The American player's goals are those of the Kentucky troops in the 19th century – a need to obtain scalps. To do so, an American unit needs to spend the turn following a melee collecting trophies.

For the Native American player, gaining social status and material goods is a key concern. One attains these by capturing prisoners and other war trophies.

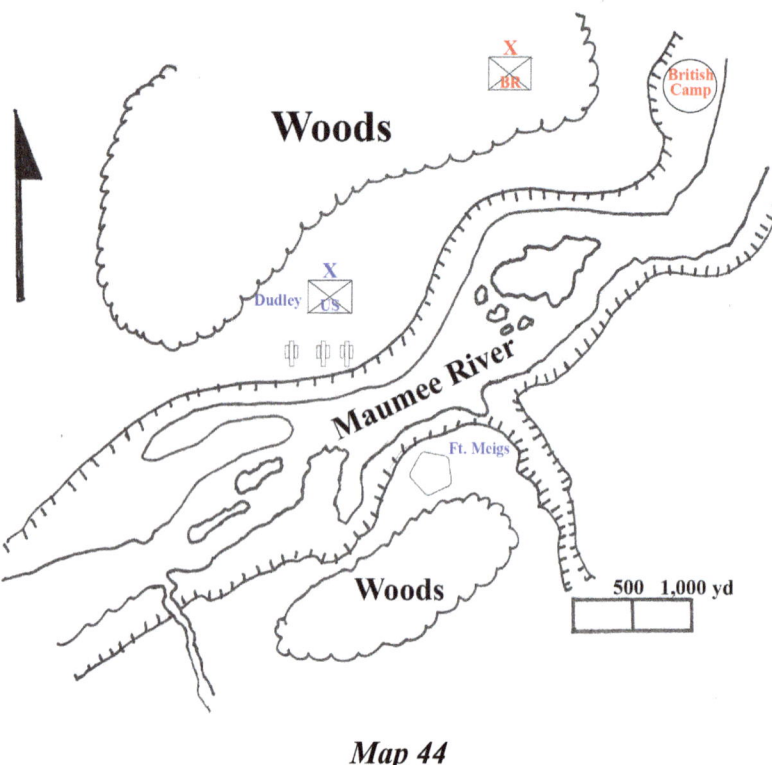

Map 44

Scenario Order of Battle
British - 860 men
Royal Newfoundland Regiment, one company, 63 men
Native Americans, 800 men

US - 860 men
Lt. Col. William Dudley
800 men, ten companies
Twenty-Fourth Infantry Regiment, one company, 60 men

Scenario Victory Conditions
The American player is at a disadvantage, as the gamer has a harder time controlling the troops. The company of the Royal Newfoundland Regiment is considered to be disordered at the start of the scenario. Both sides gain a victory point by collecting war trophies – one point is awarded for every two casualties. The side with the most trophies wins. The game plays for twelve turns.

Scenario #2 – Clay's advance to the fort
Brig. Gen. Green Clay had followed Harrison's orders and sent the larger portion of his force to attack the four British artillery batteries along the north side of the Maumee River. Clay was unable to bring his boat to the north shore and along with four other boats, landed on the south side of the river a mile west of the fort. The landing did not go unnoticed and the Americans took fire from both Native Americans and British guns. Faced with stiff resistance, Harrison ordered the dragoons and a company of infantry to help Clay. Faced with attacks on both sides, the Native

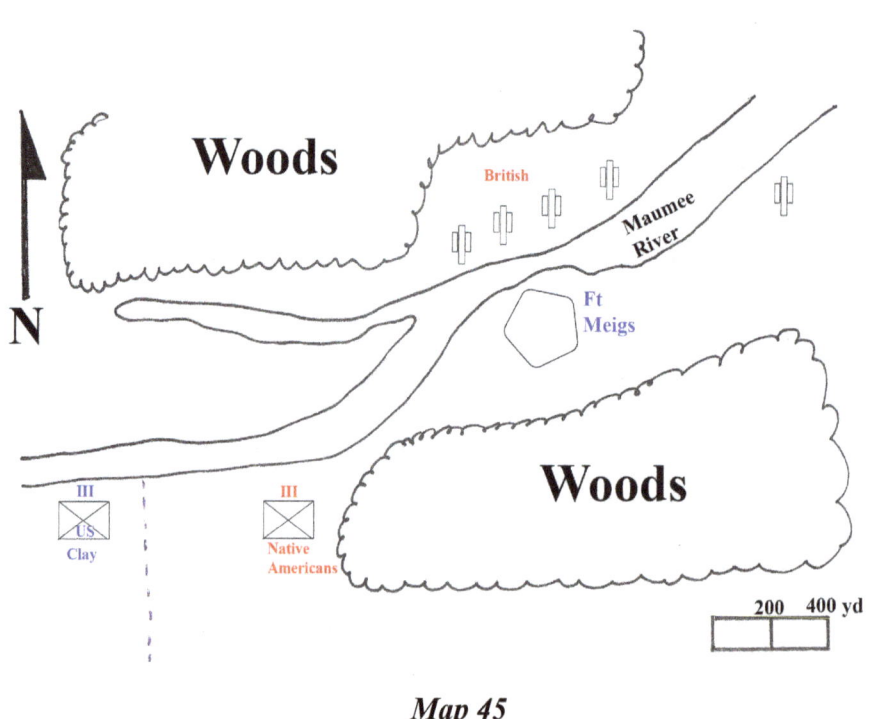

Map 45

Americans gave way and Clay was able to reach the fort.

Scenario Order of Battle
British - 600 men
600 Native warriors

US - 700 men
At the Boats
Brig. Gen. Green Clay
Lt. Col. William Boswell
400 men

At the Fort
Dragoons, 200 men
Nineteenth Infantry Regiment, one company, 60 men

103. Outside the walls of Ft. Meigs, courtesy of Harold Oney.

The game begins with the American boats have landed on the shore and are still within the boats. The boats were modified by the crews before reaching Ft. Meigs to provide cover against small arms. It will take a turn for an American company to get out of a boat.

A die roll determines when the American reinforcements can move out of the fort. Starting at the end of turn 1, a roll of a 6 on a D6 will make these troops available at the start of the following turn. If the US player does not roll a 6, at the end of turn 2, a 5 or a 6 is needed to release the troops. The required die roll will continue to decrease by 1 until the figures appear on the table. Even with the author's tendency to fail every die roll, the American reinforcements will appear on the table at the start of turn 7.

Clay can ride a horse but will need to spend a turn next to a dragoon to mount the animal. Once he is mounted, Clay then gains the additional move distance.

Scenario Victory Conditions
Clay needs to reach the fort and he has ten turns to do so. If he fails to be within a move of the fort, the British automatically win.

Scenario #3 – American attack on the British Battery
Harrison knew that the placement of the British battery on the southern side of the river was the largest threat to the fort and launched an attack against it. When the British realized the Americans had attacked the battery, they launched a counterattack but by this point the damage had already been done. With the gun taken out of action, Procter's best chance to take the object had evaporated.

Historically, the American raiding force slid out of the fort and assembled in a ravine just outside the stockade. From there they advanced up toward the small hill and at fifty yards, fired on the battery. As the Americans then charged the battery, Native Americans emerged from the woods and began to fire. Counter charged by the 41st Foot, the Americans were able to spike the guns and retreat back to the safety of the fort, taking with them a number of prisoners.

When recreating this game on the table, the American player is quick to recognize their distinct disadvantage in numbers. However, they do have surprise on their side, as the Native Americans and the flank companies of the 41st only become active when the attackers begin to shoot. If, or when, the Americans troops become outflanked, an additional US Infantry Company is released from the fort.

To spike the guns, a unit must spend 1 turn at a gun to destroy it. No dice are needed to complete the task.

Scenario Order of Battle
British - 500 men
At the Battery
Militia, two companies, 120 men
One 12 or 24-pdr. gun
One 5.5-inch mortar or howitzer
In the neighboring woods
Native Americans, 240 men

In reserve
Flank Companies, 41st Foot, 120 men

US 350 men
Commanding Officer: Col. John Miller
Seventeenth and Nineteenth US Infantry Regiments, seven companies, 240 men
Volunteer Battalion
Petersburg Volunteers, 64 men
    Pittsburg Blues, 25 men
    Greensburg Riflemen, 12 men

In reserve
US Infantry, one company, 60 men

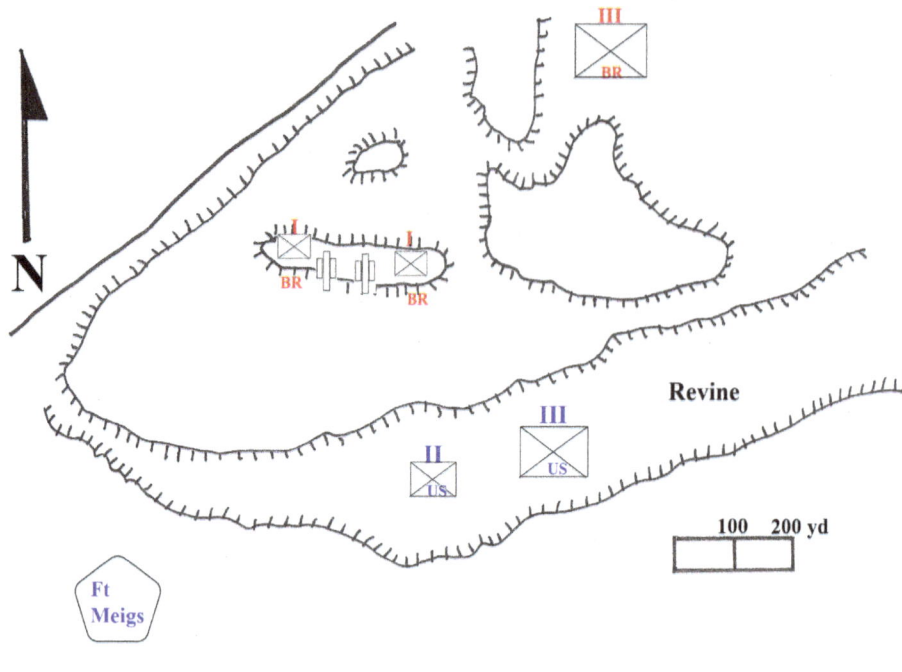

Map 46

Victory Conditions
    The Americans need to spike the guns. The British need to stop this from occurring. The game plays for ten turns.

Scenario #4 – Catch a Cannonball
    The British had arrived at Ft. Meigs well stocked with ammunition and supplies. The Americans, however, lacked cannonballs for most of their guns and relied upon the British to provide them with round shot. Once the British had fired a cannon ball, the Americans would scramble after the ball and return it to the commissary for a ¼ jig of whiskey. Apparently, over a 1,000 cannon balls were recovered by the Americans in the days before Clay's troops arrived at the fort.
    This game is best played with a small group of friends on a small table. One figure for each player works best. Take twenty markers (small rocks, dice, or something similar) and place ten of them evenly spaced along two sides of the playing. Roll two differently colored D10s – the resulting die roll indicates were a cannonball lands. Each player then moves their figure the maximum move rate to the area with the impacted ball. If a figure reaches the cannon ball, they can pick it up. If two or more players reach the cannonball, they each roll a D6 – the higher die roll indicates who has possession of the ball. The following turn they will need to bring the ball to the commissary, where they are given a drink for their troubles. The commissary is in an underground bombproof and is represented in the game by a small wooden building in the center of the table. The commissary cannot be destroyed. If a figure has returned a ball to the commissary, they cannot be hit by a cannonball when they are getting their drink. Only 1 ball can be carried at a time.
    The next turn, two more cannonballs are dropped on the gaming space. Each player can move toward

whichever ball is closest. If a figure is at the spot where the ball drops, the figure has a 50% chance of becoming a casualty. A 4, 5, or 6 on a D6 die roll indicates the figure survives and will be in a good position to retrieve the ball. For the remaining 8 turns, 3 additional cannonballs are fired at the fort.

Scenario Specifics
- A handful of figures to represent the American troops within the fort
- Twenty small rocks or similar markers to provide grid locations on the gaming table
- Cannonball marks. 24 markers in total will be used in the game, but as they can be reused, 12 could work.
- A jig or shot glass
- A bottle of your favorite beverage

Scenario Victory Conditions
The game plays for ten turns or until the bottle containing your beverage of choice is empty.

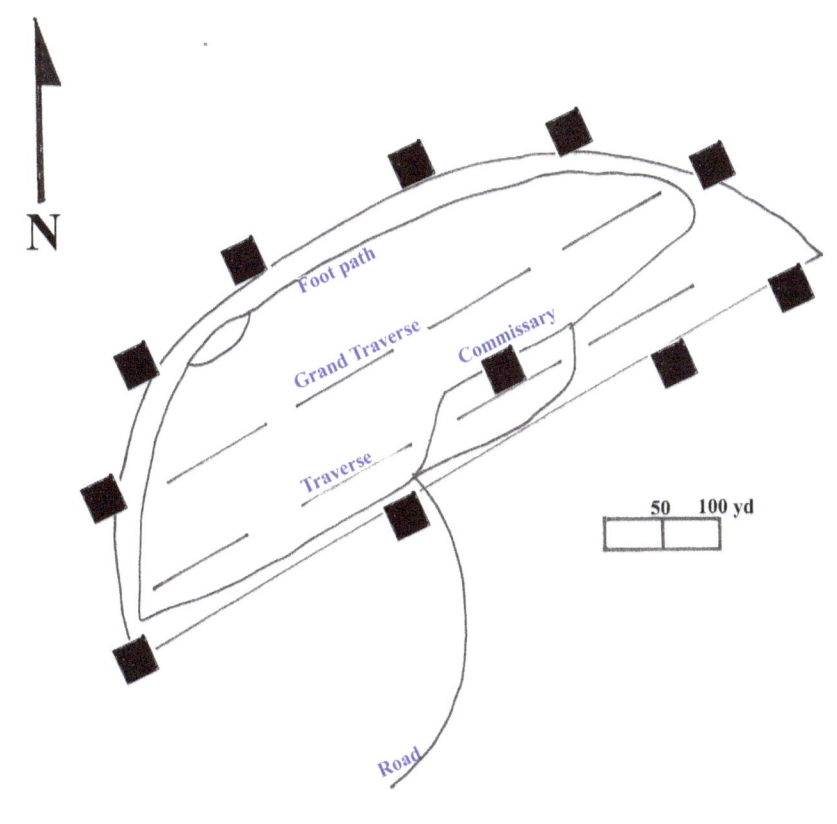

Map 47

Further Reading

Emch, James L., *All But Forgotten*. (Trafford Publishing. Bloomington, IN. 2013)

McAfee, Robert B., *History of the Late War in the Western Country*. (Worsley & Smith. Lexington KY. 1816)

Nelson, Larry, Dudley's Defeat and the Relief of Ft. Meigs during the War of 1812. *The Register of the Kentucky Historical Society*, Winter 2006, Vol. 194, No. 1., 5-42. 2006

104. Frontier Militia in Hunting Shirts, courtesy of Sash and Saber Castings.

Put-In-Bay, 10 September 1813

Overview
Naval Engagement

The American and British Navies battle on Lake Erie.

Introduction
At the beginning of the 19th century, there were few armed ships on the Great Lakes. The term lake, however, is misleading, as they are fresh-water seas. Lake Superior is the largest freshwater body of water in the world and all five of them are large enough to effect weather patterns. Given the role water routes played in both shipping and settlements, it is not surprising that when the European moved into the area, they settled around the shores of the Great Lakes. Separated by the Falls at Niagara, the bodies of water were split into two separate theaters – 1) Lake Ontario, which connected with the St. Lawrence River and then the Atlantic and 2) the remaining four Great Lakes: Lakes Erie, Huron, Superior and Michigan.

At the start of the war, there were few warships on the four western lakes. Great Britain maintained one 17-gun ship, the sloop-of-war *Queen Charlotte,* and the US the six-gun brig *Adams*. Both sides increased the number of their vessels by purchasing civilian craft and converting them into makeshift warships. When Gen. Hull surrendered Detroit to the British, the American *Adams* was taken by the Provincial Marine and put into service as the HMS *Detroit*. The *Detroit*, along with the sloop-of-war *Queen Charlotte* and schooner *Lady Prevost* gave the British temporary naval superiority on Lake Erie. Though the US managed to recapture the *Detroit*, the vessel was burnt rather than allowing it to fall into British hands once again.

105. Oliver Hazard Perry by Jane Stuart circa 1857, Birmingham Museum of Art.

For any American advance to take place in the western Great Lakes, the Americans needed to establish control on Lake Erie. The Americans began to build ships at Presque Isle at Erie Pennsylvania and Black Rock, New York. At Black Rock, Lt. Jesse Elliot, USN, purchased three civilian schooners and the recently captured *Caledonia* and began to convert them into warships. At the new shipyard at Presque Isle, Daniel Dobbin began building two warships, which was later expanded to include two brigs.

The British noted the American building activity but were unable to match the American output. The United States sent skilled shipwrights and sailors, while the British sent few men and supplies. By the spring of 1813, the Americans had launched two new gunboats on Lake Erie. Following the American assault on Ft. George, Master Commandant Oliver Hazard Perry was able to sail five warships from Black Rock - the brig *Caledonia*, schooners *Ohio*, *Somers*, *Amelia*, and the sloop *Trippe* - to Presque Isle.

Perry's flotilla now consisted of eleven vessels, but as most seamen were sent to Lake Ontario, the US naval presence on Lake Erie lacked trained Sailors. Perry's other issue was the two new brigs drew too much water to cross a shallow sand bar that protected the harbor. While Perry could lighten the brigs by removing their guns in order to cross the sand bar and enter Lake Erie, the vessels would be unarmed if a British squadron arrived during this procedure.

In June, Commodore Robert H. Barkley, RN arrived with two dozen Royal Navy sailors and a few officers to take command of the British naval forces on Lake Erie. Recognizing the difficulty Perry would have in getting his brigs over the sand bar, Barkley placed Presque Isle under a blockade while focusing his attention on getting his newest ship completed. The British departed to replenish their supplies and Perry began getting his two brigs into the lake. With brigs out in the lake, though not yet re-armed, the British returned. Perry formed a battleline which included the unarmed brigs and the British, not having a boat that could stand up to even one brig, sailed away.

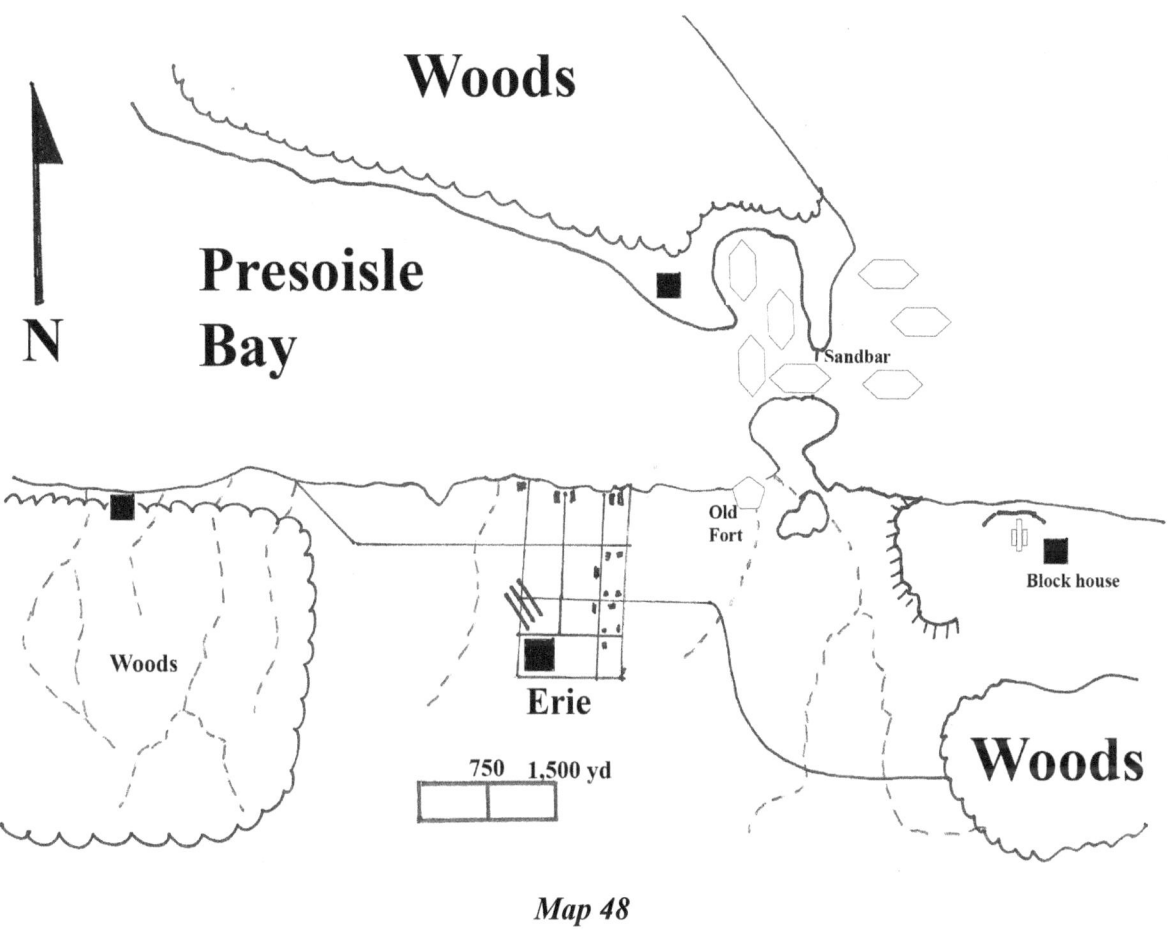

Map 48

Once Perry's brigs were re-armed, the Americans dominated the lake. Perry made contact with Ft. Meigs and was able to press into service the men he needed to operate his ships. The British, on the other hand, were blockaded at Ft. Malden and quickly running out of rations. With the launch of the newly completed 20-gun sloop HMS *Detroit*, Barkley laid up the *Erie* in an effort to crew his flotilla. On 9 September, the British outclassed flotilla left the Detroit River and sailed into Lake Erie.

The Battle

Warned of the British advance, Perry made his craft ready for battle. With a light wind to his back, Perry's plan was for the *Lawrence* and the *Niagara* to close with the enemy, while his lighter boats would attack the enemy from longer range. Both squadrons were deployed in a line of battle. With a special blue and white battle flag carrying the words "Don't give up the ship", Perry's flagship charged the British line. The *Detroit* was the first to fire on the Americans – its first 24-pdr. ball missed the *Lawrence*, but its second shot was on target. Passing into close range, the *Lawrence* demolished the *Detroit*, but as the *Niagara* failed to advance, Perry's flagship became surrounded by the *Queen Charlotte* and the *Caledonia*. With the *Lawrence* reduced to only one working gun, Perry transferred his command to the *Niagara* and proceeded to demolish the previously damaged *Queen Charlotte*. Thereafter the remaining British ships struck their colors.

In the course of three hours, including the half hour for Perry to transfer his command from the *Lawrence* to the *Niagara*, Perry scored a major victory. Perry's message to Harrison, "We have met the enemy and they are ours" signaled not only a British withdrawal from Ft. Malden, but thay Lake Erie would remain under US control for the rest of the war.

Order of Battle
British
Commodore Robert H. Barkley, RN
HMS *Detroit,* frigate, one 18-pdr. on swivel, two 24-pdrs., six 12-pdrs., eight 9-pdrs., one 24-pdr. carronade, one 18-pdr. carronade
HMS *Queen Charlotte,* frigate, one 12-pdr., two 9-pdrs., twelve 24-pdr. carronades
HMS *Lady Provost,* schooner, one 9-pdr., two 6-pdrs., ten 12-pdr. carronades
HMS *General Hunter,* brig, four 6-pdrs., two 4-pdrs., two 2-pdrs., two 12-pdr. carronades
HMS *Chippawa,* schooner, one 9-pdr. long gun
HMS *Little Belt,* sloop, one 12-pdr., two 6-pdrs.

US
Master Commandant Oliver Hazard Perry
USS *Lawrence,* brig, two 12-pdrs., eighteen 32-pdr. carronades
USS *Niagara,* brig, two 12-pdrs., eighteen 32-pdr. carronades
USS *Caledonia,* brig, two 24-pdrs., one 32-pdr. carronade
USS *Scorpion,* schooner, one 32-pdr., one 32-pdr. carronade
USS *Ariel*, schooner, four 12-pdrs.
USS *Somers,* schooner, one 24-pdr., one 32-pdr. carronade
USS *Tigress,* schooner, one 32-pdr.
USS *Trippe,* sloop, one 24-pdr.

106. Battle of Lake Erie, C. W. Jefferys.

Game Notes

There are two possible wargame scenarios which can be drawn from this naval action. The first is when Barkley runs across the unarmed brigs in Perry's battleline and the second is the actual naval engagement which was fought on 13 September 1813. Both actions can make for great tabletop recreations.

In recreating Perry's bluff, both the *Niagara* and the *Lawrence* are stripped of their cannons. They are considered to be partially manned and will strike their colors when the hulls or rigging suffer over 50% damage. If one is using Warlord's Black Seas rules, both brigs have a 7 break value.

In recreating the events of 13 September, the British are in a do or die situation and must sweep the

Americans from the lake. The *Detroit* is bigger than anything the Americans have afloat and Perry's flotilla is drastically outgunned at long range. The American squadron has more guns and crew, but the failure of the *Niagara* to join the battle almost resulted in a victory for the Crown. Though the commander of the *Niagara*, Capt. Elliott, later claimed he did not want to break the line of battle. To reflect Elliott's reluctance to commit, the American player needs to roll D6. On a roll of a 1 or a 2, the *Niagara* moves only half the distance moved by the *Lawrence*. Once the *Lawrence* is destroyed and Perry transfers his flag to the *Niagara*, the ship can move as normal. Perry can only leave the *Lawrence* when the ship has no sails and no guns.

107. The restored U.S. Brig Niagara demonstrates 1813 era naval gunnery, photo by John Baker, courtesy of the Erie Maritime Museum.

Victory Conditions

This is a fight to the death. The British must destroy the American flotilla and the Americans must destroy the British squadron.

Further Reading

Altoff, Gerry, *Oliver Hazard Perry and the Battle of Lake Erie*. (The Perry Group. Put-in-Bay, Ohio. 1990)

Elting, John R., *Amateurs, To Arms!: A Military History of the War of 1812*. (Da Capo Press. New York. 1995)

Lardas, Mark, *Great Lakes Warships 1812-1815*. (Osprey Publishing. Oxford. 2012)

Lossing, Benson J., *Pictorial Field-Book of the War of 1812*. (Harper& Brothers Publishing. New York. 1869)

108. We have met the enemy and they are ours ..., courtesy of Chris Ngiau.

Thames, 5 October 1813

Overview
Large Battle - approximately 5,000 participants

US troops destroy a retreating British force and decimate the Native Americans fighting for the Crown.

Introduction

Following the September 10th American naval victory on Lake Erie, the British position on the entire western peninsula of Upper Canada was no longer defendable. The British forces consumed large quantities of both flour and cattle, neither of which could be sourced locally and needed to be shipped in from the east. Perry's naval victory at Put-In-Bay provided the US with the ability of either cutting the British supply line, and thus starving the British and their Native American allies, or directly attacking the British and defeating them at their leisure.

Recognizing that the Americans had gain the initiative, the British operating along the Detroit area began to retreat eastward. Withdrawing from Ft. Malden, the British commander Brig. Gen. Henry Procter knew that the strength of the army relied upon the continued Native American cooperation. For many of the Native Americans, the memory of British betrayal - once at the end of the American War of Independence and more recently when they sought sanctuary at Ft. Miami after the Battle of Fallen Timbers – remain fresh. Procter failed to communicate his planned retreat to his Native American allies, which caused further discord between the two allies. Some Native Americans chose not to retreat with the British Army, while others, believing that cooperation with the British as their best option for stopping American settlement, fell in with the British retreat.

109. Battle of the Thames, C. W. Jefferys.

On 24 September, Procter left Malden and began his retreat up the Thames River. The American army led by Maj. Gen. William Henry Harrison, was slow to follow, as he waited for Johnson's Mounted Kentucky Infantry to join him. The Kentucky militia under Harrison was poorly trained and only recently armed, while his one US Infantry Regiment was too small to take on the British directly. The Kentucky Mounted Infantry under the command of Richard Mentor Johnson were well drilled and eager to fight the Native Americans.

When Harrison finally launched his pursuit, the British had already spent a week pushing their supply boats up the Thames River. Procter had told Tecumseh that the British would make a stand at Chatham. By 2 October, the American advance guard had caught up to the British. Though Procter had planned for a defensive battery to be constructed at Chatham, none was built and following a small skirmish, the British continued their retreat eastward. More Native Americans, disillusioned by the British failed promise to make a stand at Chatham, left the British.

The British had abandoned many of their supplies and had sunk their boats in an attempt to block the American gunboats from proceeding further up the Thames River. Refugee wagons packed the roads and further confused the British retreat. Moraviantown, a settlement of less than a hundred houses, was filled with Native American and civilian refugees. Further retreat would result in abandoning the refugees to the ravages of war.

Rather than fight in the town itself, Procter placed his men a mile southwest of the settlement and waited for the Americans.

The Battle

Procter had reconnoitered the ground previously and had placed his one remaining 6-pdr. cannon on the main road. British troops were deployed on both sides of the guns while Tecumseh and men loyal to his cause occupied the woods on the northern flank. Tecumseh himself took up a position between the British and men loyal to his cause. A few Canadian Dragoons were employed as pickets. For the next two or three hours, they waited for the Americans to show. The British regulars had spread out, as to make themselves a smaller target for the Americans.

The original American assault called for Harrison's regulars to push on the British line, while the Kentucky Mounted Troops were held in reserve. Johnson, seeing the weakness of the British position, asked and received permission to charge the British. The Kentucky Mounted Troops charged the gun successfully and finding themselves to the rear of the British line, dismounted and began to fire on the enemy. The British regulars, subject to fire from both sides, routed from the battlefield. The flight of the 41st exposed the Native American flank, which was attacked in turn by the Americans. Johnson himself wounded, killed the Native American chief who shot him.

Tecumseh was killed early in the battle, as various retreating British troops later attested to seeing him lying dead on the battlefield. Though some controversy remains as to who killed Tecumseh, his death, along with the rout of the British regulars, impacted the resolve of the remaining Native American warriors. The remaining Native American warriors disengaged from the fight and some drifted back to the British lines. Unpursued, the remains of the British Army limped back to Burlington Heights while the Americans were satisfied in capturing and sacking Moraviantown.

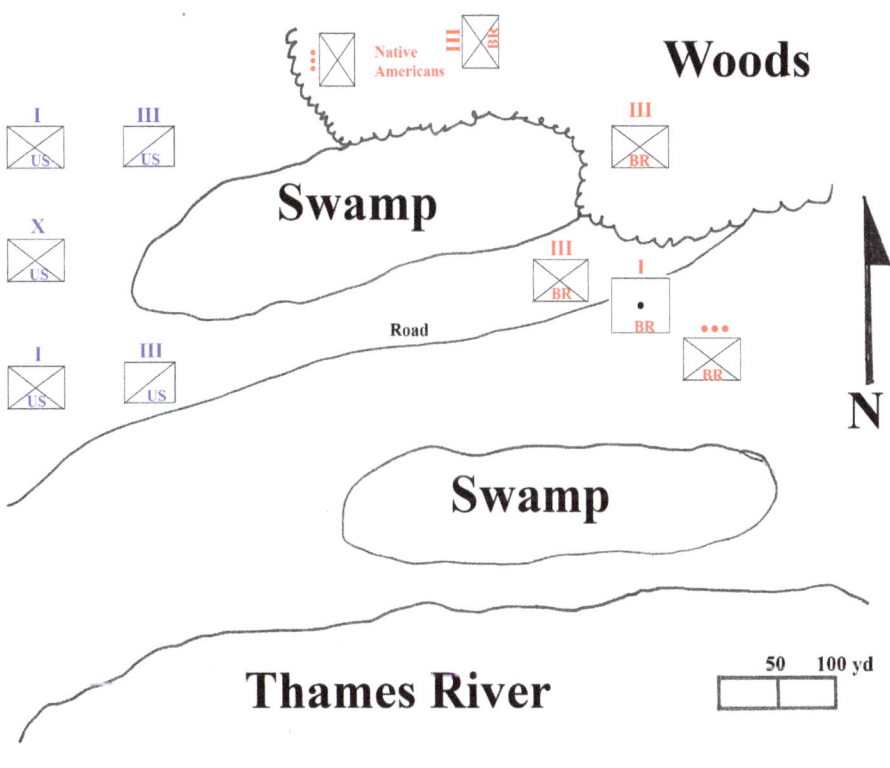

Map 49

Order of Battle
British - 1,300 men
Commanding Officer: Maj. Gen. Henry Procter

41st Regiment of Foot, 400 men
Canadian Light Dragoons, 20 men
10th Royal Veteran Battalion, 40 men
Royal Artillery, 30 men, one 6-pdr. cannon

Tecumseh, 800 men

US - 3,000
Commanding Officer: Maj. Gen. William Henry Harrison

Twenty-Seventh US Infantry Regiment, 120 men
Mounted Kentucky Militia, Col. Richard Mentor Johnson, 1,000 men
Kentucky Infantry (foot), 1,800 men

110. Recaptured 3-pdr cannon, author's photo.

Game Notes

Procter's conduct during the campaign resulted in a court-marshal. Suspended of rank and pay for six months, Procter returned to England and was retired from British service. Conversely, Harrison and Johnson both personally benefitted from their actions – Harrison was elected the 9th president and Johnson served as vice-president under Martin Van Buren.

Tabletop wargaming has the benefit of 20/20 hindsight and given what is known about the British loss, a British player is unlikely to make the same errors that Procter did. The British troops, for one, would not be spread out but placed closer together to avoid being overrun by cavalry. Secondly, given the time between when the British first occupied the ground and the Americans arrived, a competent British officer would have improved his defenses. Though the British had most likely been lacking shovels for digging entrenchments, the British troops were versed in the construction of abatis - cut down trees with the branches pointing out toward the enemy. Had the British constructed an abatis, Col. Johnson would have been unable to launch his mounted charge.

Popular accounts of the battle were the mainstay of mid-19th century American mythology and by most accounts, the battle was almost exclusively fought by the Mounted Kentucky Militia against the combined British and Native American forces. To this day the roll of the US regulars and Kentucky non-mounted militia remains unclear and for those wanting to recreate the action of the tabletop, a more balanced game would limit the engagement to the Native Americans, British regulars, and mounted Kentucky troops.

111. Kentucky Mounted Infantry, courtesy of Knuckleduster Miniatures.

In our recreation of the encounter, the British and Native Americans occupy the same ground and initially only the mounted Kentucky troops take part. In our ten-turn game, additional American reinforcements appear at the start of turn six. For the Americans to win, they must drive the British from the field. To keep things moving, one victory point is given to each unit destroyed in the course of the game and two victory points are given for the death or capture of an enemy commander. In this recreation, the American mounted volunteers are divided into twelve groups, while the British have one artillery unit, four units of regulars and six of Native Americans. The British also have two commanders (Tecumseh and Procter) and the Americans have two (Col. Richard Mentor Johnson and his brother Lt. Col. James Johnson). At the start of turn 6, Harrison, five units of Kentucky Militia and one unit of regular US troops appear at the edge of the table.

Victory Conditions

The Americans must destroy BOTH the British army AND the Native Americans to win the game. The British win the game if they hold the field. The game plays for ten turns.

Further Reading

Elting, John R., *Amateurs, to Arms! A Military History of the War of 1812*. (Da Capo Press. New York. 1995)

Hitsman, J. Mackay, *The Incredible War of 1812*. (Robin Brass Studio. Montreal. 2012)

Sugden, John, *Tecumseh's Last Stand*. (University of Oklahoma Press. Norman. 1985)

112. British.Line, courtesy of Pendraken

Mackinac Island, 4 August 1814

Overview
Small Battle - approximately 1,000 participants

An American force of regulars and militia tried to recapture Mackinac Island. Landing on an undefended side of the island, the US troops marched inland to try to capture the weaker side of the fortifications. Aware of the American advance, the British pulled their troops from the two forts and brought their forces to the high ground of their choosing. As the initial American advance stalled, US regulars attempted to turn a flank, but were ambushed by a group of Native Americans. Unable to advance further, the US troops retreat to their boats.

Introduction
At the start of the war, a small British force of veterans and Native Americans captured Fort Mackinac on Mackinac Island. Following the American victory on Lake Erie, the US sought to recapture northern Michigan. Aware of the fort's vulnerability from an overland attack, the British built a stockade blockhouse, later to be known as Ft. George. In late July, the US flotilla approached the island, but it became clear that the big 32-pdr. could not be elevated sufficiently to bombard the British held fortifications.

On 4 August, the American forces landed on the west side of the island and proceeded to march inland to attack Fort Mackinac on its weakest side. The British commander, Lt. Col. Robert McDouall was aware of the American plan of attack. Leaving behind a small token force to guard the two forts, McDouall brought his men to an open farm field on the northern end of the island. The British troops held the high ground on the south side of the field, which was strengthened by breastworks. Native Americans working with the British were posted in the woods on both flanks.

113. Ft Mackinac (Lossing).

The Battle
As the Americans emerged from the woods on the north side of the farm field, the two British field guns opened up. The Americans withdrew back into the wood line. Forming the men into two lines, with the militia in front, the Americans once again emerged from the woods. The Americans brought up their own artillery, two 6-pdrs., but their fire on the British was ineffective. The Americans pressed forward and tried to encircle the British forces. Hearing that the Americans had landed more troops closer to the British forts, the British commander pulled some of this troops back from the line. Most of the Native troops rushed back, hoping to ambush the Americans as they were landing. Some Natives remained at their post, however, and when US regulars under Maj. Holmes sought to turn the British flank, a small group of Menominee Natives opened fire. Holmes died instantly and the attack stalled. Without a leader, the Americans fell back in confusion. Realizing that no Americans were landing at the fort, British troops began to return to the fight. Unable to turn the flanks and insufficient in number to directly assault the entrenched British, the Americans retreated back to their boats.

Order of Battle
British
Commanding Officer: Lt. Col. Robert McDouall

10th Royal Veteran Battalion, 20 men
Royal Regiment of Artillery, 2 NCOs, 11 men, one bronze 6-pdr. cannon, one bronze 3-pdr. gun
Royal Newfoundland Fencible Infantry, 100 men
Michigan Fencibles, 37 men
Militia, 2 companies, 100 men – although one company of 50 men were left guarding the forts)
Natives, 350 warriors of various nations

US
Commanding Officer: Lt. Col. George Croghan, Second Rifle Regiment

Maj. Andrew Holmes, Thirty-Second Infantry Regiment, commanding the US regulars
Five companies of US regulars from the Seventeenth, Nineteenth, and Twenty-Fourth Infantry Regiments, 500 men (1 company remained with the landing boats – 400 men in the assault)
Col. Cotgreave's Regiment of Ohio Volunteers, 250 men
Detachment of Regular US Artillery, two 6-pdr. guns
US Marines, a detachment that stayed with the landing boats

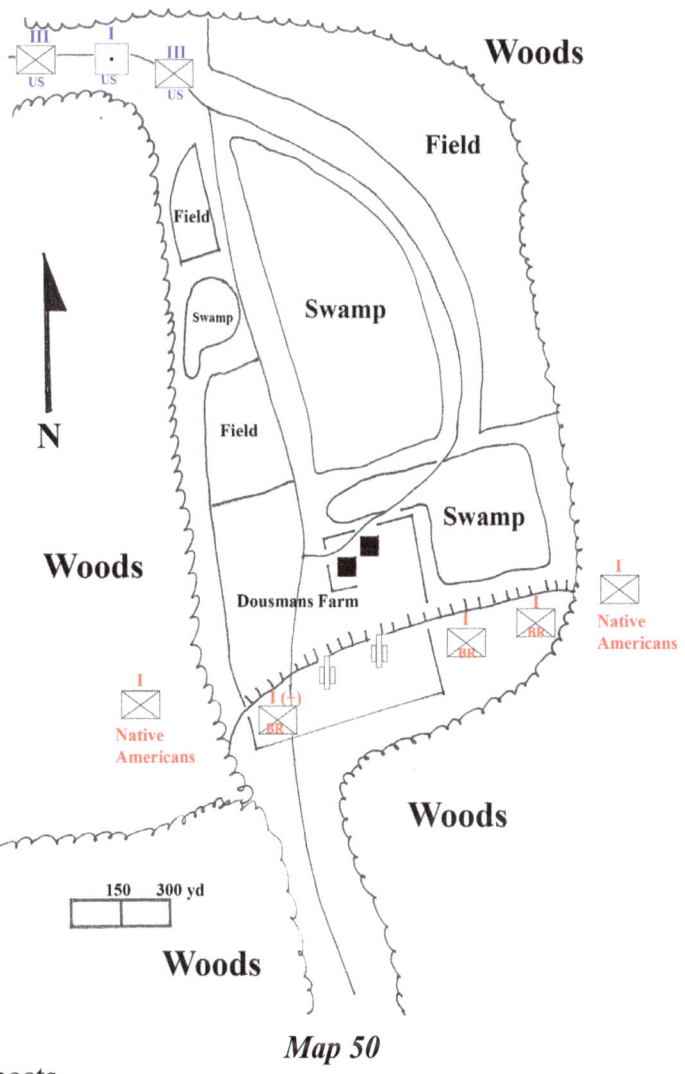

Map 50

Game Notes

The American force, while strong on paper, was ill-prepared to retake the island. As the US boats were unable to elevate their guns sufficiently, the planned direct assault on the island was wisely discarded. Disembarking at the spot where the British came ashore a few years previous was rather predictable, and the naval pre-bombardment of the landing site removed all doubt as to the American intentions. The plodding US advance allowed time for the British to pick the ground on which to engage the Americans and make best use of their Native allies. Had the British been overpowered, they could always retire to their fortifications and try again. As Native warriors were unlikely to hole up in the two forts, they could have launched an attack on the rear of the Americans.

The major question remains - could the US troops have taken the island, even after recognizing that their guns were unable to batter the two forts directly? Had the Americans disembarked only part of their force, they could have landed their remaining troops near the forts and taken them, as they were held by only a token force. Unfortunately for the Americans, Croghan, who was best known for defending a glorified blockhouse (Siege of Ft. Stephenson), went with a simple plan, hoping that the Natives would get tired of waiting in the two forts and abandon their British allies.

When gaming this scenario, the US player should be allowed to keep part of his force in reserve to attack the British forts. After launching the US advance of the British line, the British commander will learn that the Americans have landed to the rear. He then can order British units to the rear, but each Native warband needs to roll to see if they stay in position or fall back to the forts. Conversely, the Natives may move to attack the Americans directly, if they see a good opening. For game purposes, the British player can roll a die roll to see if they can attack the Americans directly. If the Natives fail to launch an attack, they will leave the battlefield.

114. 40mm US Infantry, author's photo.

The following modifiers are suggested in making the die roll on a D6. For each applicable modifier, an additional pip is added to the outcome. If the British player fails to gain a total of 6 points when attempting to launch the Natives against the Americans, the warband leaves the table
- Americans are facing away from the Natives
- An American unit is outnumbered by a Native group
- The American unit has suffered more than 50 casualties
- An American senior officer is dead

When gaming with Native Americans, any rules should reflect the operational mentality of the indigenous population. Natives were not simply local recruits raised to fill up the ranks of a White Man's army but were the product of their own societies with their own views of honor and bravery. Two hundred years of interactions with Europeans had prepared them to participate in the conflict and they acted in their own self interests. By the early 19th century, Natives had been using firearms for generations, so they should not suffer any penalty in using them. At the same time, Native troops did not form ranks and did not go toe to toe in prolonged firefights. Since each member of the war party knew each other since childhood, they sought to minimize any casualties. If an attack appeared not to be going in their favor, Natives saw no dishonor of leaving the fight.

How does this translate to the tabletop? First off, the Natives will not attack a strong fortified position directly and will not try to besiege a location for more than a few days. Most Europeans and colonists were afraid of the Natives and when meeting them in the woods they often panicked. Other colonists and regular troops stood up against the Natives and fought well against them. In a melee, especially after springing an ambush, Natives should get an additional bonus, but they were unlikely to stay around if they did not break their foe after the first round.

For game purposes, the Americans have two battalions (1 militia and 1 regulars) and a two-gun battery. The American player can split off companies from each battalion to form a detached skirmish unit and can divide his battery into two one-gun units.

The British have one battalion, a two-gun battery, and two companies of militia under their direct command. The British player can place the warbands into position at the start of the battle but should not control their actions after the fighting starts. Sources disagree to the number of Natives at the battle, but for game purposes, you can split them into 6 warbands (the number of figures in each warband is up to the British player). The scenario plays well at a scale of about 1 to 10. The 1 to 10 ration works well for regular troops, but each two-gun battery has a few extra figures. In our games, each gun has a gun model and 4 gunners.

Victory Conditions

Major British Victory:	Drive the Americans off the field, kill or capture a senior American officer
Minor British Victory:	Keep the Americans from carrying the British entrenchments
Minor US Victory:	Drive the British from the field without destroying 2 or more British units
Major US Victory:	Drive the British from the field AND destroy 2 or more British units.

Further Reading

Dunnigan, Brian, *The British Army at Mackinac.* Reports in Mackinac History and Archaeology, Number 7, Mackinac State Historic Parks. 1980

Elting, John R., *Amateurs, to Arms! A Military History of the War of 1812*. (Da Capo Press. New York. 1995)

May, George S., *War 1812.* (Mackinac State Historic Parks. 1970)

115. British troops, courtesy of Tod Kershner.

East Coast

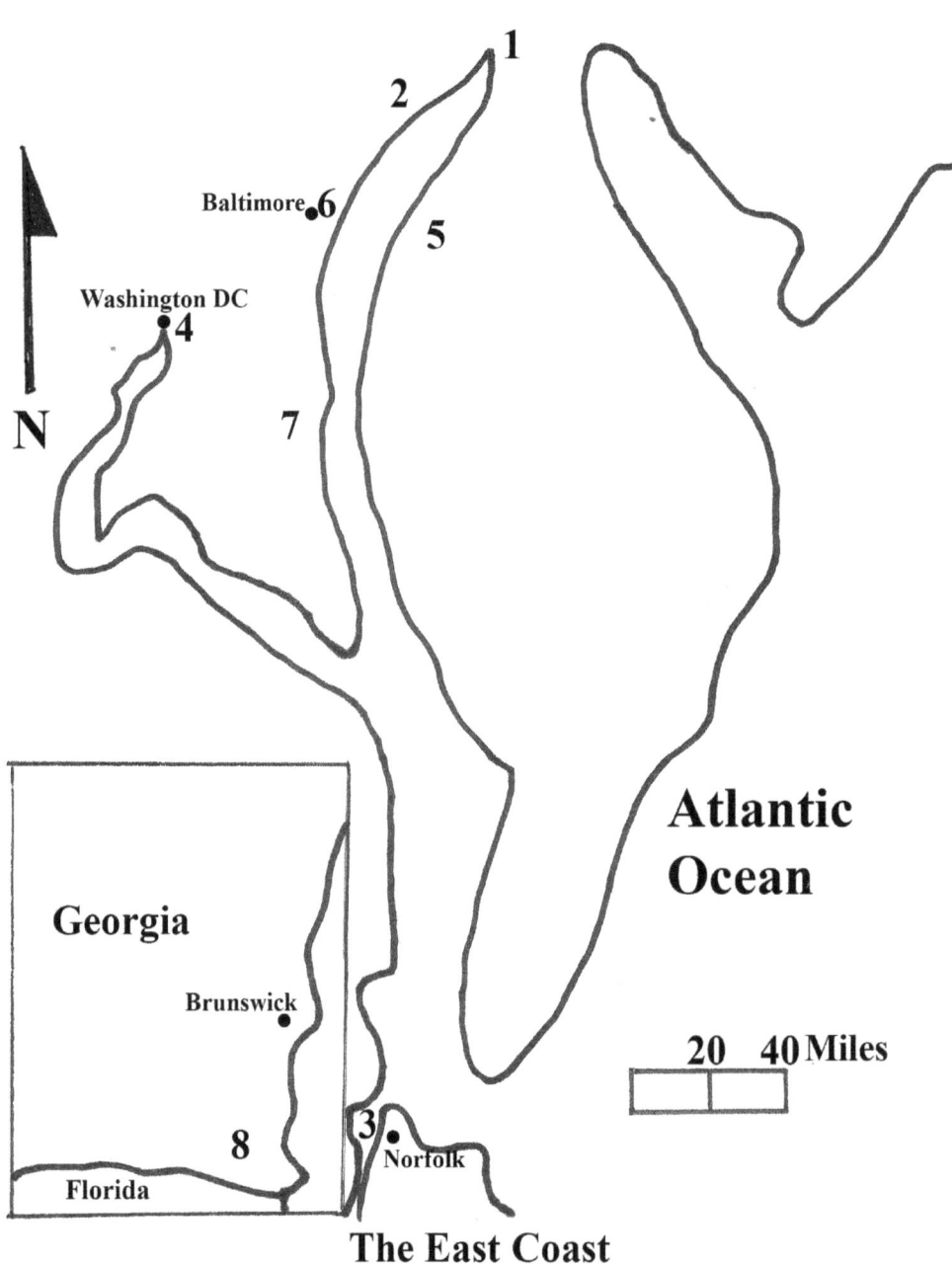

The East Coast

1. Frenchtown 2. Harve de Grace 3. Craney Island 4. Bladensburg
5. Caulk's Field 6. North Point 7. Kirby's Windmill 8. Point Petre

Map 51

Raid on French Town, 29 April 1813

Adrian O. Mandzy

Overview
Small Action - less than 200 participants

British raid on the Hamlet of French Town.

Introduction

With the majority of the British forces committed to the struggle against Napoleon in Continental Europe, few troops could be sent to Upper Canada. In order to draw American troops from the Northwest frontier, as well as to disrupt local shipbuilding and military stores, the British dispatched Rear Admiral George Cockburn to the Chesapeake. A career officer, Cockburn despised the Americans and allowed his troops to personally enrich themselves at the expense of the civilian population. Arriving in the area in early 1813, Cockburn would spend the next two years terrorizing the Chesapeake.

The US Secretary of War John Armstrong was unwilling to defend what US President James Madison referred to as "ever man's turnip patch" in Maryland and left its defense to its Governor, Levin Winder. With US regular troops confined to a handful of forts guarding the larger naval yards, the smaller settlements were left to defend for themselves.

Once they arrived in the Chesapeake area, the British first tried to capture or destroy the US Navy frigate *Constellation*. Following the failed attempt to capture the *USS Constellation* on 20 March, Cockburn led his forces up the bay to chart the region's waters. Along the way, the British captured a number of fast American schooners as they worked their way up to the northern side of the Chesapeake.

The hamlet of Frenchtown was located approximately thirteen miles up the Elk River and by 1813 was a prosperous commercial port. Goods manufactured in Baltimore and destined for American troops in Canada passed through the settlement. A semi-completed earthen battery, mounting four brass 6-pdrs., protected the river channel at the Lower Warf Landing. Learning of the close proximity to the British, the Americans hastily called out the local militia, while cattle and personal property were quickly moved to the backcountry. Once the British learned that the settlement had a depot of flour and other military stores, it was only a matter of time before they decided to pay Frenchtown a visit.

The Battle

At around 7:00 in the morning, eleven armed British barges under Lt. George Westphal approached the town. The number of British landing barges is unknown, but they were armed with 12-pdr. carronades. Following a brief exchange of cannon fire, the British shore party landed. In face of the enemy, the local American defenders fled, abandoning the town. The Royal Marines proceeded to burn the tavern, two storehouses, stables and the two schooners anchored at a small wharf.

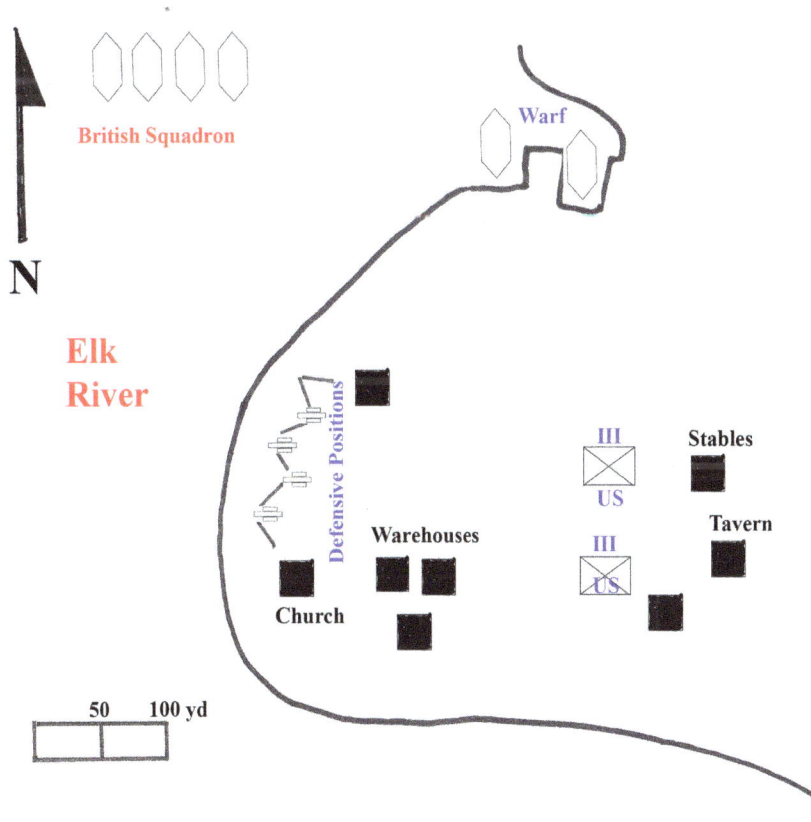

Map 52

116. British Landing, courtesy of Perry Miniatures.

Order of Battle
British
Commanding Officer: Lt. George Westphal

Royal Marines, 100 men
> For game purposes divided into five groups of Royal Marines

US
Commanding Officer: Maj. James Sewall

Local militia, 60 men
> For game purposes divided into four groups of gun crews and two groups of musket armed men.

Game Notes
The British have a distinct advantage in both troop quantity and quality. However, the British troops must first land and it is unclear if they will be able to do so all at the same time. The Americans also have four guns that they are able to fire at the British and an earthen battery to hide behind. The British gunboats, though equipped with cannon, cannot elevate to directly hit the men behind the earthen redoubt while the guns of the British squadron are too far away to accurately hit the battery.

Victory Conditions
The British need to capture the town and set it aflame. For every building set afire, the British gain a victory point. The Americans need to defend the town and gain a victory point for every two Royal Marine casualties. The player with the highest number of victory points wins. The game plays for ten turns

117. Militia in Civilian Dress, courtesy of Sash and Saber Castings.

Further Reading

George, Christopher T.
2000   *Terror on the Chesapeake: The War of 1812 on the Bay.* White Mane Books. Shippensburg, PA.

Quick, Stanley L.
2015   *Lion in the Bay: The British Invasion of the Chesapeake.* Naval Institute Press. Annapolis

118. Royal Marines, courtesy of Richard Watts.

Havre de Grace, 3 May 1813

Overview
Small skirmish - less than 200 participants

Americans flee from a British raiding force.

Introduction
    The town of Havre de Grace was located where the Susquehanna River flowed into the Chesapeake. Named after a city in France, the town was an important stop on the Post Road between Baltimore and Philadelphia. Given the settlement's location along existing trade routes, Samuel Hughes had started an iron foundry northeast of the settlement. As early as the 1790s, the factory had been casting iron cannon for the Navy. To protect the town, an earthen three-gun battery was erected at Havre de Grace.

    Rear Admiral George Cockburn's distain of the Americans was well known and following his earlier raids on Frenchtown, Elk Landing, and Spesutie Island, Cockburn turned his attention on Havre de Grace. The admiral noted an American flag flying over one of the town's two batteries and targeted the area for destruction. Once the defiant Americans had fired on the British flotilla from their shore battery, Havre de Grace would face the wrath of the Royal Navy.

The Battle
    Cockburn, having previously scouted the shallows of the Chesapeake, noted that the water level was too low for his larger ships to pass and embarked his raiders on smaller barges. Departing the safety of the flotilla around midnight, the four assault groups reached the shoreline undetected. Thought the Americans were well aware that the British were in the area, no pickets or patrols were posted. The American commander, Lt. Col. William Smith, was not in the area at the time of the attack and only a skeleton crew remained at the main battery.

    At dawn's first light, the British attacked Havre de Grace. After a brief exchange of gunfire, a Congreve rocket struck an American defender in the head, killing him instantly. Witnessing his death, the remaining American militia skedaddled to the town, where they continued to take pot shots at the British. At the smaller secondary battery Lt. John O'Neil remained at his post and continued to fire at the British.

    Once the British captured the main battery, they turned the guns on the town. The British then launched a bayonet charge that dislodged the remaining militia from Havre de Grace. As the militia were fleeing to the woods on the edge of town, Cockburn landed and gave the order for his troops to begin looting. Horses were butchered in their stalls and hogs were cut open and allowed to run free. Afterwards, the troops set fire to the town. Property damage to the town was to exceed $50,000 in 1813 prices.

    While in Havre de Grace, Cockburn learned of a warehouse at the upriver ferry crossing and of the iron factory nearby and sent two raiding parties out to destroy these facilities. Neither place mounted any serious defense against the British. The warehouse at Smith Ferry was put to

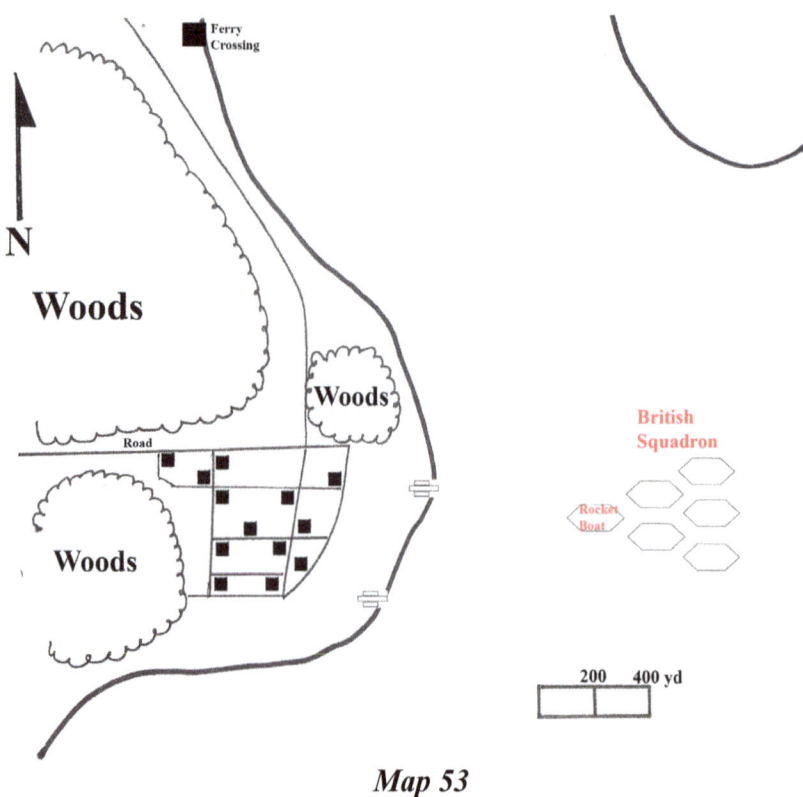

Map 53

the torch, as were other building. At the factory, a five 24-pdr. gun battery protected the iron works, but it was not manned nor ready for action. The British troops spent the rest of the day destroying the factory machinery, along with 46 cannons. Satisfied with the destruction, the British returned to their boats.

Order of Battle
British
Commanding Officer: Rear Admiral George Cockburn

Royal Marines, 150 men
Royal Artillery, 20 men
11 landing barges
One rocket boat

US
At Havre de Grace
    Main Battery
        Courtney's Company, 98 men, Capt. Thomas Courtney
    Secondary Battery
        Whiteford's Company, 71 men, Capt. William Whiteford

At the Ironworks
    Five 24-pdr. cannons, unmanned*.
        * Had these been crewed the British forces would have been decimated.

On Route to Annapolis
    Fortieth Regiment Maryland Militia
    Forty-Second Regiment Maryland Militia

119. Rear-Admiral Sir George Cockburn by John James Halls circa 1817, National Maritime Museum.

Game Notes
    Cockburn launched his attack without a good understanding of the area and got lucky that he did not meet a prepared enemy. The American Militia was mostly disbursed at the time of the attack and the battery at the ironworks was not crewed. The failure of the Americans to post patrols and pickets at Havre de Grace was not unique, as was the absence of their commanding officer. While the result of this raid only served to reinforce Cockburn's low opinion of citizen participation, his grace would always be facing an unprepared enemy.

    The raid of May 3rd has three distinct phases – the attack on Havre de Grace and the two follow up attacks on the ferry and the ironworks. If one were to play out the raid with only the American forces that took part, it would be a very one-sided victory. To mitigate the one-sided nature of the event, we suggest introducing additional American reinforcements. Not only should the militia at the two batteries be at full strength, but their commander should be present as well. Furthermore, additional American troops including militia cavalry and up to two regiments of US Maryland Militia may appear at Havre de Grace or at the two other targets.

    Clearly, the game requires the British to press their attacks. Cockburn only learned of the iron works once he had secured the town, so his force first needs to control Havre de Grace for one turn before he can set out for the ironworks. Once that is accomplished, he needs to send off two assault groups to destroy the other targets while a third force needs to secure the town from a possible American counterattack. Historically, Cockburn divided his command into four groups, but the British player can split the troops differently.

Bad Roads and Poor Rations

120. Battlefield monument at Havre De Grace, author's photo.

The game plays for twelve turns. At the start of Turn 6, the American rolls to see if any reinforcements appear. A resulting 6 on a roll of a D6 has a unit appearing on a table edge. A second die roll reveals which troops appear:

1 a company of local militia volunteers, 60 men
2 a company of militia riflemen, 60 men
3 a company of trained militia, 60 men, and artillerymen, 20 men
4 a troop of dragoons, 60 men
5 the 42nd Regiment of Maryland Militia, 240 men
6 the 40th Regiment of Maryland Militia, 300 men

To determine on which edge the troops appear, a third die is rolled. On a 1 or a 2, the troops on the eastern edge, on a 3 or 4, the troops are on the northern side, a die roll of a 5 or a 6 results in the troops appearing on the western table edge. On the following turn, turn 7, a D6 is rolled again and now on a 5 or 6, a second die roll determines which troops appear. Use the same procedure to determine where the troops appear. This procedure is followed until the game ends.

Note: the same troops cannot appear twice during the game. For example, if on turn 6 and 9 American reinforcements appear and a 4 on the second die roll appears twice, only one unit of dragoons will take part in the game. The American player does not gain a re-roll.

Victory Conditions

The British gain two victory points for taking and holding Havre de Grace. The British can gain an additional victory by destroying the ferry, but the true prize is the ironworks. The destruction on the cannons and the American ability to produce more of them is something worth boasting about. Cockburn later wrote that his destruction of the iron foundry was of natural importance and for game purposes, we assign three victory points to the works.

Further Reading

Cockburn, George, The London Gazette, https://www.thegazette.co.uk/London/issue/16750/page/1331. 1813

George, Christopher T., *Terror on the Chesapeake: The War of 1812 on the Bay.* (White Mane Books. Shippensburg, PA. 2000)

Sheads, Scott S., *The Chesapeake Campaigns 1813-1815.* (Osprey Publishing. Oxford. 2014)

Painted by John Morris (Mystic Spirals)

121 Sailors with pistols, courtesy of Perry Miniatures.

Adrian O. Mandzy

Craney Island, 20 June 1813

Overview
Battle - 3,000 participants

Joint British operation to capture the city of Norfolk and destroy the *USS Constellation*.

Introduction

Having successfully attacked American settlements along the northern section of the Chesapeake earlier in the year, British returned to the Bay with additional reinforcements. The target of their raid was the city of Norfolk and USS *Constellation*. The *Constellation*, along with twenty gunboats, had been bottled up at Norfolk for much of the war, but maintaining a blockage allowed the American privateers operating from other ports in the south to prey upon the British ships operating in the Caribbean sea-lanes. Once the British eradicated the American frigates, they could focus their attention on protecting their merchantmen.

The Americans were aware of the vulnerability of Norfolk and in March 1813 Virginia Governor James Barbour call up the militia to reinforce the already stationed American troops. US Brig. Gen. Robert B. Taylor recognized the importance of Craney Island as a choke point and fortified the area using US Marines, Navy seamen, and guns from the *Constellation*. By the time of the British incursion, the fortifications on Craney Island developed into a mile-long earthwork with a seven-gun battery in the southwest part and an outer battery in the northwest. Further upriver, the Americans positioned their twenty gunboats which could provide supporting fire. The *Constellation* itself was five miles further inland and positioned between Ft. Nelson on the west side of the river and Ft. Norfolk, on the river's east side.

The Battle

As the British squadron made its way up the Chesapeake and anchored in Hampton Roads, the Americans launched a pre-emptive night attack with their gunboats on the frigate HMS *Junon*. As the Americans began their attack, the fog lifted and the wind picked up, allowing three nearby British frigates to come to the aid of the *Junon*. After two hours of firing, the Americans withdrew to shallow waters.

By the 21st, the British had successfully maneuvered their twenty-two warships and sixty barges just outside the Elizabeth River. Landing at daybreak to the west of the island, 900 British soldiers and marines moved toward the fort. A combination of high tide and accurate American artillery fire halted the first British assault.

At 1100, the British launched a second wave of thirty to forty barges onto the northeast part of the island, while the already landed British

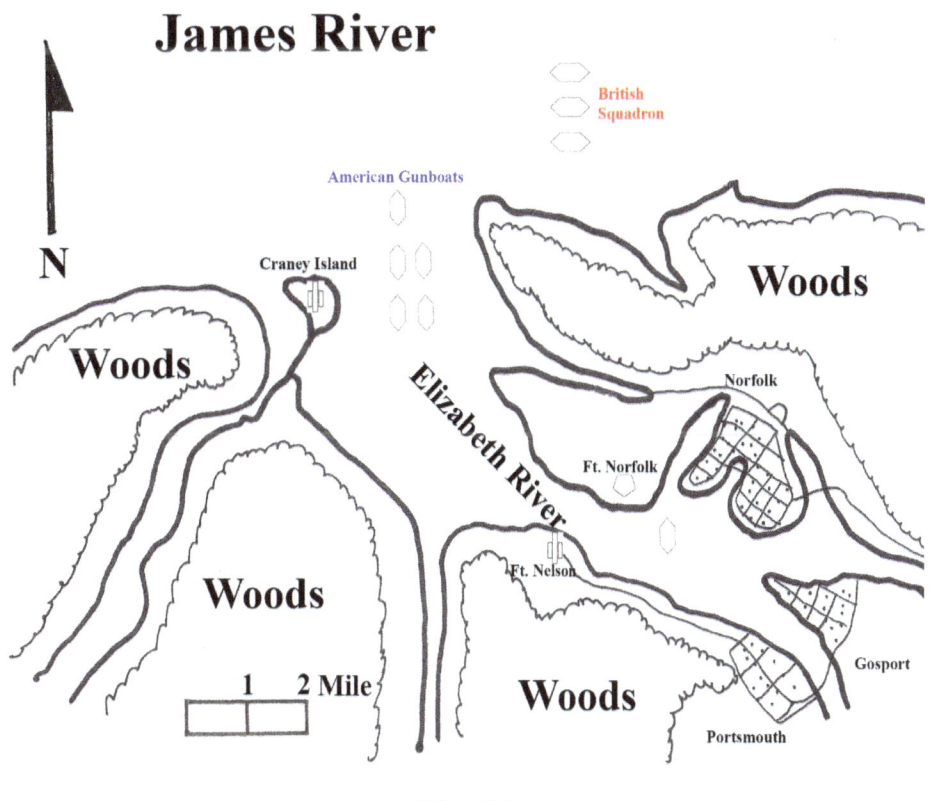

Map 54

troops renewed their attack on the western part of the fortifications. The Royal Marines fired their Congreve rockets, which were met with accurate American artillery. Close to shore, the lead British barge, the *Centipede*, was hulled by a cannonball and other boats were severely damaged. The remaining British barges retreated, and the Americans captured the Admiral Warren's ornate *Centipede*.

Order of Battle
British
Commanding Officer: Admiral Sir John B. Warren
Capt. Samuel J. Pechell, RN

British Army Forces
Col. Thomas Sydney Beckwith, RA

Twenty-two warships, including: six ships of the line (HMS *Marlborough, Victorious, Diadem,* and *San Domingo*), six frigates (HMS *Junon, Narcissus,* and *Barrosa),* and three brigs (HMS *Fantome, Mohawk* and *Contest).* 60 landing barges

Landing Party, 2,350 men
Lt. Col. Thomas Sydney Beckwith
Lt. Col. Charles James Napier
Royal Marines, two battalions, 1,642 men
Royal Marines Artillery, two detachments, *100 men* (two 6-pdrs.)
Royal Marine Rocket Artillery, half of a company, 50 Men
102nd Regiment of Foot, 300 men
Independent Foreigners, two companies, 250 men
Royal Navy Landing Party, *50 men*

US
Commanding Officer: Brig. Gen. Robert Barrund Taylor

On Craney Island, *800 men*
Brig. Gen. Robert Barrund Taylor
39 US Marines and 111 Navy Seamen from the USS *Constellation*
4th Virginia Regiment (militia) under Lt. Col. Henry Betty, *450 men*

122. British Flat boat, courtesy of Perry Miniatures.

Maj. Andrew Waggoner
Capt. Thomas Robert's Rifle Co. (31st Regiment), *40 men*
Twentieth US Infantry under Lt. Johnson, *30 men*
Capt. Samuel Shields Rifle Co., *50 men*
Virginia Militia (artillerymen), *80 men*
First battery, 18-pdrs on the North West part of the Island
Second battery, two 24-pdrs. and four 6-pdrs.

At the Mouth of the Elizabeth River
Master Commandant Joseph Tarbell, USN
20 gunboats

Ft. Nelson and Ft. Norfolk
At the forts *100 men*
Anchored between the forts was the USS *Constellation* with a reduced crew

123. USS *Constellation*, courtesy of GHQ Models.

Game Notes

There are various ways one can recreate this battle on the tabletop. While one can game it as an Infantry assault, the battle is more of a ship and shore operation. Using Warlord's Age of Sail Battle Game *Black Seas*, one can recreate all of the components of the battle on a 6x10 table.

The American earthen fort has a 360-degree arc of fire. Being a stable gun platform, all rules should allow an additional modifier. Manned by experienced gunners, the gunners at the fort were firing at known ranges, so they should receive a +1 to hit modifier.

To cross the water between the western edge of the island and the mainland, the British player needs to roll a D6 on every turn. If the British player fails to roll a 6 on their first attempt to cross, they must remain on the other side of the water for the rest of the turn. After two turns, the British player may roll again, but this time needing a 5 or a 6 on a D6. If any British troops are still in position to cross after 4 turns have passed, they need only to roll a 4, 5, or 6 on this or any subsequent turns.

If the British are able to capture the fort, any remaining US gunboats may attack the landing craft. The gunboats themselves were a product of the Jefferson administration. With a shallow draft these vessels were incapable of operating in heavy seas and needed to stay close to shore. The gunboats were armed with a heavy cannon, a 32 or 24-pdr. on a traversing carriage and were propelled by sail or oars. At close range, the heavy long cannon would have been potentially devastating to the British landing barges.

The British squadron could fire upon the American gunboats, but as they would have been unable to move closer, their fire would be at long range. The *Constellation* had been stripped of its Sailors and Marines, but the cannons of Ft. Nelson and Ft. Norfolk were trained to meet the British assault.

Victory Conditions

The British have ten turns to capture the fort and if successful, ten additional turns to capture the *Constellation*.

Further Reading

Lossing, Benson J., *Pictorial Field-Book of the War of 1812*. (Harper& Brothers Publishing. New York. 1869)
Sheads, Scott S., *The Chesapeake Campaigns 1813-1815*. (Osprey Publishing. Oxford. 2014)
Simmons, Edwin H., The Battles of Craney Island and Hampton. *Fortitudine: Newsletter of the Marine Corps Historical Program.* Vol. XV, No. 2. Fall 1985. 3-7. 1985

Bladensburg, 25 August 1814

Overview
Major Battle - more than 5,000 participants

British march on Washington and the Americans try to stop them at Bladensburg.

124. Joshua Barney, (Lossing).

Introduction
In terms of military targets, US Secretary of War John Armstrong believed Washington, D.C. to be a poor choice. Besides having a few imposing Federal buildings, Washington was a sleepy southern town of no strategic value. Neighboring Baltimore, however, with its "nest of privateers" offered the British a much more tempting target.

Warned repeatedly of the coming peril, the US administration created the Tenth Military District, which included Maryland, Northern Virginia and the District of Columbia. Command of the Tenth was entrusted to Brig. Gen. William Winder, whose previous military experience amounted to him being captured in the 1813 Battle of Stoney Creek. Recently paroled, Winder was appointed to this position directly by President Madison on account that his uncle was the governor of Maryland.

Winder recognized the immensity of the task ahead of him but failed to properly prepare. With less than five hundred US regulars garrisoning the regions forts, Winder would need to call on the militia to defend the region. US Secretary of War Armstrong advised that calling out the militia too early was not feasible, as the men may not turn out and they would be reluctant to remain under arms for long periods of time when no immediate threat was apparent. Lacking a proper staff, Winder was unable to organize an effective outpost and courier system to inform him of the actions of the British squadron. Winder also failed to properly reconnoiter the terrain and identify the best defensive ground to stop the British. Most importantly, Winder was negligent in establishing mustering places for the militia to rally.

On 1 August 1814, British Adm. Sir Alexander Forester Inglis Cochrane's force of twenty warships and thirty transports departed Bermuda. With him were Maj. Gen. Robert Ross and his Peninsula War veterans. Once Rear Admiral Sir George Cockburn had disbursed the US gunboat flotilla guarding the Chesapeake, the region was secure for the British to begin landing troops. With no cavalry to scout ahead and lacking artillery cannons, Maj. Gen. Ross's goal was to launch a series of powerful raids to draw American troops away from Canada. Accompanied by Rear Admiral Sir George Cockburn, who had conducted similar raids all along the Chesapeake the previous year, the British were well prepared. On 18 August, the British sailed up the Patuxent River and landed unopposed the following morning at the town of Benedict. From there the British marched to Upper Marlboro, as from this location they could strike at either Washington or Baltimore.

News of the British landing evoked a panic in the capital and civilians began to flee the city, clogging the roads with wagons. In response to this growing threat, President Madison called out 6,000 Maryland Troops to defend Washington, of which 2,000 responded. More troops dribbled in in batches and Brig. Gen. Stansbury's Baltimore Militia encamped along the Bladensburg Road and awaited orders. President Madison appeared in front of the troops and after giving a speech, issued additional orders. The British launched a feint towards Woodyard, which resulted in the already exhausted American militia marching and then counter-marching. In dealing with the unknown, Winder scattered his forces in an attempt to defend everything.

On 22 August, Stansbury's Militia had begun digging artillery entrenchments and fortifying Lowndes Hill, a position which dominated Bladensburg. Inexplicably, Winder ordered Stansbury to retreat westward across the Anacostia River. Stanbury complied with the order and marched his exhausted men westward, throwing away a major American tactical advantage.

By Tuesday, 23 August, Winder began to concentrate all troops at Bladensburg. Taking positions on the western heights beyond the river, the exhausted American Troops continued to be assigned positions by Winder, US Secretary of State James Monroe and US Secretary of War John Armstrong, and their aides. Eventually, the Americans established a two-line defense, with a final third line to the rear. The British troops, on the other hand, rested for a day before resuming their advance on Washington.

The Battle

Observing the disposition of the Americans across the river from a two story brick building in Bladensburg, Ross noted that the two defensive lines were too far apart to support each other and launched Thornton's Light Brigade forward. American Riflemen and Artillery stopped the initial British assault, but as more British troops forded across the river and began to threaten the American flanks, the first line of defense broke. Winder led an ill-fated counterattack, but the militia, under fire from three sides and battered by Congreve rockets, broke and fled the field. As more British troops crossed the river, the second American line collapsed.

By around 1400, the British moved forward and approached the final American defensive line. Part of the British Army moved to flank the American left flank while another brigade concentrated on pushing the Americans in the center. On the flank, the Columbia Brigade stood in line, while the British engaged the Americans in light skirmish order. In time, the British regulars were able to push the Columbia Brigade and the Americans broke. In the center, defended by the remnants of the Chesapeake flotilla, a small detachment of US Marines, and three 18-pdrs., Commodore Joshua Barney's men stood their ground. When threatened by the British, Barney's men launched a successful counterattack, shouting "Board them" as they charged the enemy with boarding axes, cutlasses and pistols. Though the action held back the British assault for about half an hour, the US cannons had depleted their limited ammunition (the civilian drivers had fled the battlefield with the ammunition carts), and the defenders were eventually overwhelmed by the British.

After the defeat of Barney's men, Winder moved to Tenleytown. Though a number of American troops remained willing to fight, Winder had abandoned Washington to the enemy. Ross's men marched unopposed into Washington that evening. After firing the Capital buildings, the British seized the President's Mansion and proceeded to loot it. Cockburn himself took great delight in taking a cushion from First Lady Dolly Madison's chair, as it would remind him of "her seat". The following day, the British marched back to their boats without further American interference.

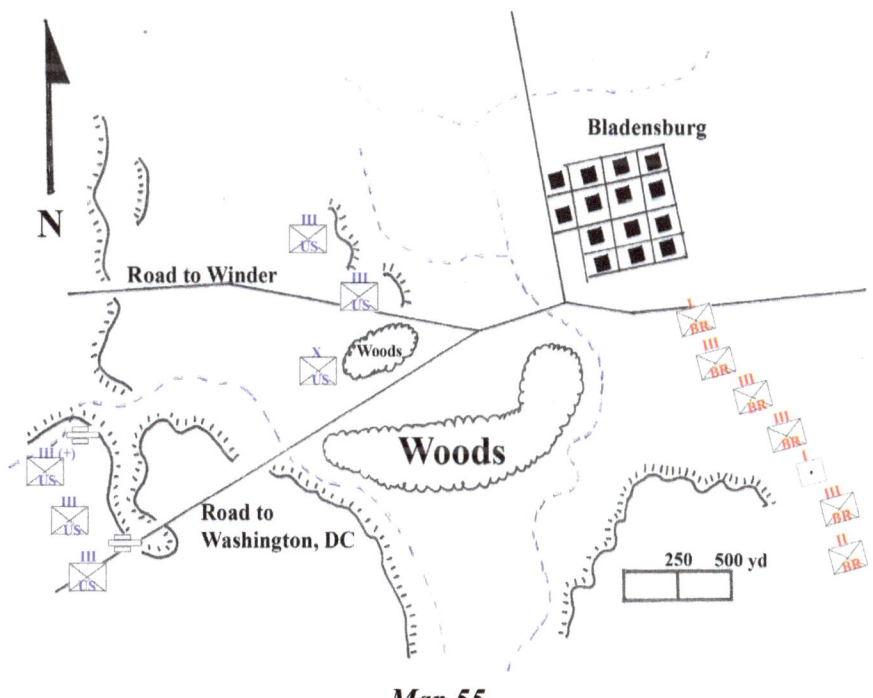

Map 55

Order of Battle

British, 4,100 men (2,800 men participated and 1,300 men held in reserve)
Commanding Officer: Maj. Gen. Robert Ross

Light Brigade, 1,100 men
Col. William Thornton
    85th Light Infantry, 600 men
    4th, 21st, and 44th Foot, light companies, 300 men
    Royal Marines, 70 men
    Colonial (West Indian) Marines, 75 men

Second Brigade
Col. Arthur Brooke, 1,300 men
    4th Foot (minus the light company)
    44th Foot (minus the light company)

125. Bladensburg, view from the British position author's photo.

Third Brigade (held in reserve and did not play a role in the battle)
Col. Patterson, 1,300 men
    21st Foot (minus the light company)
    Royal Marines, 2nd Battalion (minus the one company with the Light Brigade)

Royal Artillery, Rocket Brigade, 30 rocket tubes

Royal Navy
Rear Admiral George Cockburn
Officers and Sailors, 247 men
Royal Marines, Rocket Corps, 30 rocket tubes

US 6,000 men
Commanding Officer: Brig. Gen. William Winder

At the first two Lines
Brig. Gen. Stansbury
Fifth Maryland Regiment, 500 men
First Maryland Rifle Battalion, 150 men
First Maryland Artillery Regiment, six 6-pdrs.

In a Ravine
Second US Light Dragoons, mounted, 55 men
Lt. Col. John Tayloe, Columbia District Cavalry, 120 men
Lt. Col. Frisby Tilghman, First Maryland Cavalry District, 300 men

At the Third Line
Brig. Gen. Walter Smith

On the American left flank
First Columbian Brigade, 1,700 men
First Regiment
Second Regiment
Twelve 6-pdrs.

In the Center
Lt. Col. William Scott, 400 men
Twelfth US Infantry Regiment
Thirty-Sixth US Infantry Regiment
Thirty-Eighth US Infantry Regiment

Commodore Joshua Barney, USN
Sailors from the Chesapeake flotilla, 400 men
Three 18-pdrs.
Capt. Samuel Miller, USMC
US Marines, 120 men

On the American right flank
Col. William Beall
Seventeenth Maryland, 500 men
Lt. Col. Hood
Thirty-Second Maryland, 400 men

In the area
Virginia Troops
Lt. Col. George Minor
Tenth Virginia Regiment, 700 men
James City Light Infantry, 100 men

126. British Light Troops skirmishing, courtesy of Perry Miniatures.

Game Notes

The Battle is hard to win for an American player. The British have better quality troops and a clear chain of command. The troops are predominantly seasoned veterans, as are their officers. Both are used to the vigor's of campaign and to do what is necessary to win. Though lacking in cavalry for reconnaissance or to launch a charge on the retreated enemy, the British made do. The American cavalry was newly mounted and as raw recruits, were never used in the battle. More serious was the lack of British field artillery and they were forced to rely upon the notoriously inaccurate Congreve rockets.

Congreve rockets were used by the British in a number of battles and there are records of the enemy being killed by the technology at more than one occasion. More importantly, the rockets caused chaos among horses and untrained troops. For game purposes, rocket fire is resolved like regular artillery fire, but to determine if a rocket actually hit its target, a secondary die roll is required – in this case an unmodified 6 on a D6 is needed. If a 6 is rolled, an additional die roll determines the number of casualties the target unit suffers. On a roll of a 1 or a 2, the unit suffers 1 casualty, on a roll of a 3 or a 4, the unit takes 2 casualties, and on a 5 or 6, the unit takes 3 casualties.

In addition, any unit (both friendly and enemy) within the 180-degree firing arc of the rockets needs to take a morale check. The British troops are accustomed to the rockets and only have a slight chance of being spooked (a one in six chance – on a roll of a 1 on a D6, the

127. US Marines, courtesy of Richard Watts.

British unit drops a morale level), while the American militia and cavalry have a 50/50 chance of dropping a morale level. Every time the rockets are launched, all units within a twenty-inch range of the rockets must check their morale.

The Americans have a significant numerical advantage over the British but are hampered by their commander. By any account, the senior US commander was unfit to command a corporal's detail, much less to be entrusted with a defense of the capital. At both the strategic and tactical levels, Brig. Gen. William Winder failed.

Throughout the campaign, Winder demonstrated his inability to adequately prepare his defense. Bridges were left intact for the enemy to use at leisure. In the hours leading up to the battle, American troops were marched and counter marched, and few if any provisions were provided for them. The absence of food, ammunition, and any sort of holistic logistical support significantly hampered the American ability to resist. The presence of both Armstrong and Monroe on the battlefield did not help the American cause, as each issued often conflicting orders to the troops.

It is unlikely that any gamer could duplicate Winder's level of incompetence, so for our reconstruction, the US player needs to roll 2D6 and apply the results for every American militia and army unit on the table before the start of the game. On a roll of a

128. Royal Artillery Rocket Brigade, courtesy of Knuckleduster Miniatures.

2 - the unit moves a full move to its right
3 - the unit moves half a move to its right
4 - the unit move a full move to the rear
5 - the unit moves half a move to the rear
6 - the unit remains in position
7 - the unit remains in position
8 - the unit remains in position
9 - the unit moves half a move to the front
10 - the unit moves a full move to the front
11 - the unit moves half a move to its left
12 - the unit moves a full move to its left

The American player is also hindered by the lack of artillery ammunition. Historically, the guns in the first line were only issued round shot, which was useless when firing on skirmishers. The 18-pdrs. in the third line could only fire a few rounds and went silent as a result. To emulate this lack of ammunition, each gun battery needs to roll a D6 for the amount of ammunition it can use. The resulting die roll indicates how many times a battery can fire during a game.

The game plays for 12 turns.

Victory Conditions

The British need to drive the Americans from the field and have three units ready to exit the western table edge.

Further Reading

Eaton, H. B., Bladensburg. *Journal of the Society for Army Historical Research*, SPRING 1977, Vol. 55, No. 221 (SPRING 1977), 8-14. 1977
Elting, John R., *Amateurs, To Arms!* (Da Capo Press, Inc. NY. 1995)
George, Christopher T., *Terror on the Chesapeake: The War of 1812 on the Bay.* (White Mane Books. Shippensburg, PA. 2000)
Sheads, Scott S., *The Chesapeake Campaigns 1813-1815*. (Osprey Publishing. Oxford. 2014)

Caulk's Field, 31 August 1814

Adrian O. Mandzy

Overview
Small Skirmish - 307 participants

Nighttime British assault on an American militia camp.

Introduction

To divert attention from the British assault on Washington, Rear Admiral George Cockburn ordered twenty-nine-year-old Capt. Sir Peter Parker to lead the frigate *HMS Menelaus* into the upper Chesapeake and make some trouble for the Americans. A veteran of Trafalgar, the third-generation British naval officer was known for his dashing arrogance and propensity for the wanton destruction of property. Parker's tendency for causing mischief earned him a reputation among the Americans that rivalled Cockburn's.

In addition to drawing American troops away from Washington, Parker's mission was to collect intelligence on the areas surrounding Baltimore in preparation for the British assault. Learning from a formally enslaved man of the presence of an American militia encamped a mile from the beach, Parker reveled at the thought of having one more frolic with the Yankees. Twenty of the sailors in his command were armed with pikes for the purpose of slashing the American tents. Shortly after 20:00, Parker's men boarded their boats and sailed for the

129. Peter Parker, (Lossing).

coast. By 23:00, the British troops had landed and were marching inland. The forces of the Crown were divided into battle groups (referred to as "divisions"). Parker was riding a recently captured horse and personally leading the Royal Marines.

The Americans had mobilized the countryside for the defense of Baltimore, with one out of every five men being called up for service. Lt. Col. Reed was a veteran of the US War of Independence and had, as a young lieutenant, served under Brig. Gen. "Mad Anthony" Wayne in the elite Corp of Light Infantry. Reed took part in the nighttime bayonet charge during the 1779 Battle of Stoney Point and was acquainted with fighting at night.

The British landing was spotted by American militia dragoons, who after exchanging a few shots with the Redcoats, reported the British presence to Reed. Reed ordered his men to break camp and march to a more defendable position 300 paces to the rear. The US troops were positioned in two lines on the high ground, with their three working cannons in the center; a fourth gun was abandoned near the American camp. On the American left, a company of riflemen secured the flank by deploying in the woods slightly ahead of the first line.

Though having lost the element of surprise, the British continued their advance on to the American camp. Arriving at Caulk's field by 01:00, the British found their adversary well prepared to meet them.

The Battle

The British were marched along Georgetown Road and upon arriving at Caulk's Field, they deployed into line. The Maryland Riflemen opened fire and after the British began to advance on their position, the green coated marksmen fell back to the American line. As the British continued their advanced toward the high ground, the three American cannons began to fire. As the British approached closer, the militia began to fire as well. At about seventy paces from each other, the British returned fire. Parker, who was leading the marines in the center of the British line made a prime target, ordered his men to stop firing and move forward. For his efforts, Parker was rewarded by the Americans with a well-placed shot that severed his femoral artery.

As Parker was bleeding to death, the Americans had depleted their ammunition and began their planned withdrawal from the high ground. Though the British had captured the high ground, the death of their commander prevented the Crown forces from mounting an active pursuit of the retreating American militia. The British then retired with their wounded and reached their boat around 03:00. The following day, under a flag of truce, the British returned to collect their wounded and dead.

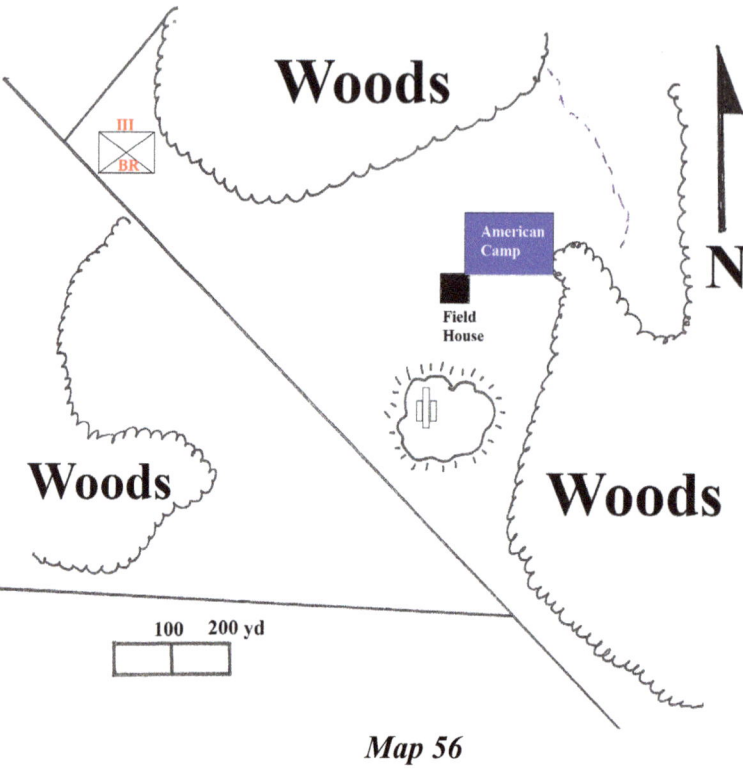

Map 56

Order of Battle
British - 150 men
Commanding Officer: Capt. Sir Peter Parker, R.N.

1st Division, 50 sailors
2nd Division, 50 sailors
Royal Marines, 50 men

US - *157 men*
Commanding Officer: Lt. Col. Philip Reed

Twenty-First Maryland Militia, 4 companies, 98 men
Rifles, one Company, 26 men
Artillery, three 6-pdr. cannons, 21 men
Militia Dragoons, *12 mounted men*

Game Notes

Having lost the element of surprise, Parker needed to reconsider his plan of attack, but his arrogance and low opinion of the Americans was typical of the British aristocracy at the time, and it would lead to his doom. On

130. Sailors attacking with boarding pikes, courtesy of Perry Miniatures.

the other hand, the American commander, Lt. Col. Reed was definitely the right person for the job – he maximized the strengths of his troops to inflict as many casualties on the British as possible. Though one could debate if Reed should have pursued the enemy as they were retiring, the American troops were low on ammunition and were in no position to press the attack.

After the battle was over, the Americans believed they had won. American newspapers glorified Reed and the actions of his men. Falling on the heels of the disastrous Battle of Bladensburg, Reed's action raised American morale and demonstrated that the British were not invincible.

As repeated in numerous more recent wars, a technically superior military force seized ground that in the end had no value. In the process of capturing this ground, the professional military suffered casualties at the hands of a militia, which in turn were then used to strengthen militia resolve in the oncoming engagements. Technically, the British had won but it was at best a Pyrrhic Victory.

In wargaming this encounter, the British need to destroy the American militia. They have better trained troops and an aggressive commander. For game purposes, the 60 men of the naval landing party are armed with muskets, while twenty have pikes and the remainder are armed with hatchets and pistols. The Royal Marines are armed with muskets and Congreve rockets, which saw no use during the encounter.

The Americans are limited in their ammunition supply (apparently the militiamen carried only 20 rounds of ammunition with them) and were armed with a variety of weapons - one post battle account describes a militiaman firing his blunderbuss which caused Parker's fatal wound. The riflemen are crack shots but have little formal training.

The American Artillerymen performed admirably during the fight and are rated as trained troops. However, they also have limited ammunition. To reflect this limitation, each gun is considered to have 2 rounds of canister shot and 1D6 rounds of solid shot. The D6 is rolled after the first shot of each gun and a corresponding number of chits are placed by the cannon by the American player. It is possible to shift chits from one cannon to another, but to do so, a gun crew must spend an entire turn bringing over as much ammunition the other gun is willing to share.

The battle was fought at night under a bright full moon. Unlike other nighttime War of 1812 engagements, the darkness did not apparently result in any friendly fire incidents. However, the darkness, combined with the lingering smoke generated by weapons firing black powder, limited target visibility and may help to explain the poor British accuracy during the engagement. The American fire was significantly more accurate and deadly – only three Americans were wounded during the action while the British reported more than forty casualties.

In our wargaming recreation of the battle, we had assumed that the British troops were silhouetted by the moonlight. The Americans, on the other hand, were holding the high ground, which made gauging target ranges more difficult. As a result, the majority of the British shots went over the heads of the Americans. To simulate this disparity, every time a British unit shoots, one of the scored hits is considered to be a miss.

The game plays for twelve turns.

Victory Conditions

The British must destroy the American force. For every American unit destroyed the British gain two victory points. The British also gain 5 victory points for each cannon they capture and ten points if they hold the high ground at the end of the game.

The Americans need to cause as many British casualties as possible. For every two British figures killed the Americans gain a victory point.

If the Americans remain on the field at the end of the game and have suffered less than 50% casualties, they automatically win the game, regardless of the number of British victory points.

Further Reading

George, Christopher T., *Terror on the Chesapeake: The War of 1812 on the Bay.* (White Mane Books. Shippensburg, PA. 2000)

Sheads, Scott S., *The Chesapeake Campaigns 1813-1815*. (Osprey Publishing. Oxford. 2014)

131. Cpt. Peter Parker and Royal Marines, courtesy of Richard Watts.

Adrian O. Mandzy

North Point, 12 September 1814

Overview

Major Battle - 5,800 participants
Large Skirmish - 800 participants
Small Skirmish - 90 participants

US land forces engage the British ahead of the American defenses surrounding Baltimore. The death of the British commander at the start of the battle results in an uncoordinated assault. This can be viewed top down from small skirmish to major battle

Introduction

Following the destruction of Washington, the British forces move to their next target, the strategically important port city of Baltimore. In 1814, Baltimore was the third largest city in the US, with a population exceeding 50,000. More importantly, the city served as a logistical center for the war effort. Privateers used Baltimore as their base of operations and the city's warehouses were full of British goods and military supplies. Well defended by artillery, Ft. McHenry, and a chain linked mast boom that blocked a 600-foot-wide channel, a direct naval assault on Baltimore was deemed unpractical.

The independent British landing on Maryland's Upper Eastern Shore was countered by a spirited American defense at Caulk's Field. During this nighttime encounter, the British Commander, Sir Peter Parker, received a mortal wound while encouraging the Royal Marines into a bayonet charge. Running low on ammunition, the Americans withdrew. The British did not press their assault and withdrew to their ship.

News of the American victory gave hope to the civilian population of Baltimore. The Americans expanded the defenses surrounding Baltimore and threw up earthworks on Hampstead Hill. On 11 September, the American troops of Brig. Gen. John Stricker's brigade moved to the North Point crossroads in anticipation of the British landing. By firing on the British and then retreating, Stricker hoped to wear down the British assault in a series of skirmishes.

The Battle

On the morning of 12 September, the British landed at North Point, ten miles below Baltimore. At the same time, the HMS *Terror* launched the first of its 190-pdr. shells onto Ft. McHenry, a star pattern earth and brick fort. Completed by 1806, the fort mounted 18-pdr., 24-pdr., and 36-pdr. cannons taken from

132. Gen. Robert Ross, courtesy of John R. Lee

a French warship damaged in 1808. American artillery fire from the fort drove back Vice Admiral Sir Alexander Cochrane's ships and prevented a more direct British assault.

Stricker deployed his militia troops in three lines on a narrow peninsula. The American flanks were protected by Bear Creek on the right and Back River on the left. To get to Baltimore, the British would need to punch a hole through the American center.

With his line of battle established, Stricker deployed Maj. Heath and 250 men as an advanced force. To get a better view of the Americans, Maj. Gen. Robert Ross, Rear Admiral George Cockburn, and a small advanced party from the 85th Foot moved to reconnoiter the ground ahead. On a grassy knoll, the mounted British officers glittered in the sunlight and made a shiny target for the First Baltimore Sharp Shooters. As Ross turned to bring up more light troops, a militiaman's ball hit its mark and the senior British general was mortally wounded.

With Ross being carried off the battlefield, Col. Arthur Brooke of the 44th Foot assumed command and pressed the American skirmish line. Deploying the 85th Foot to screen the movement of his army, Brooke called for the artillery to be brought up and opened fire on the American line. After an hour's worth of fighting the American first line was threatened. Stricker order the Thirty-Ninth and the Fifty-First Maryland up to bolster his line. The Maryland militia were unable to follow the complicated maneuver of forming a right angle and the Fifty-First promptly collapsed.

The British too were running into snags. The 4th Foot's attack stalled as it wadded into a boggy creek that fed into the Back River. Exchanging fire with the Americans, the 4th should have swung left and rolled up the American line. Instead, the British troops moved straight ahead and gave chase to the fleeing remnants of the Fifty-First.

With the British closing in on the American line, Stricker ordered his men to fall back on the defenses on Bread and Butter Creek. At about twenty yards from the American line, the British stopped their assault, which allowed the Americans to retire from the field. A final push would, in all likelihood, have broken the American resistance, but the British never launched a pursuit. The Americans were able to successfully extract their forces and reformed on the previously established defensive line, about half a mile down the road. The British, at this point, went into camp and waited to see if the Royal Navy would have better luck.

During the day, Cochrane had managed to move some of his lighter warships closer to Ft. McHenry. The fort, though somewhat battered, was able to reply with accurate and deadly force. Cochrane was forced to retreat and resume his long-range bombardment. Capt. Charles Napier of the frigate HMS *Euryalus* also tried to break into Baltimore but was forced back.

Though Brooke had reconnoitered the American defenses, he understood any further attack would have been rather costly. After conferring with Cockburn, Brooke ordered a retreat. Though an American dragoon squadron caught up to the rear guard and captured a few prisoners in a charge, the British were successful in extracting their force.

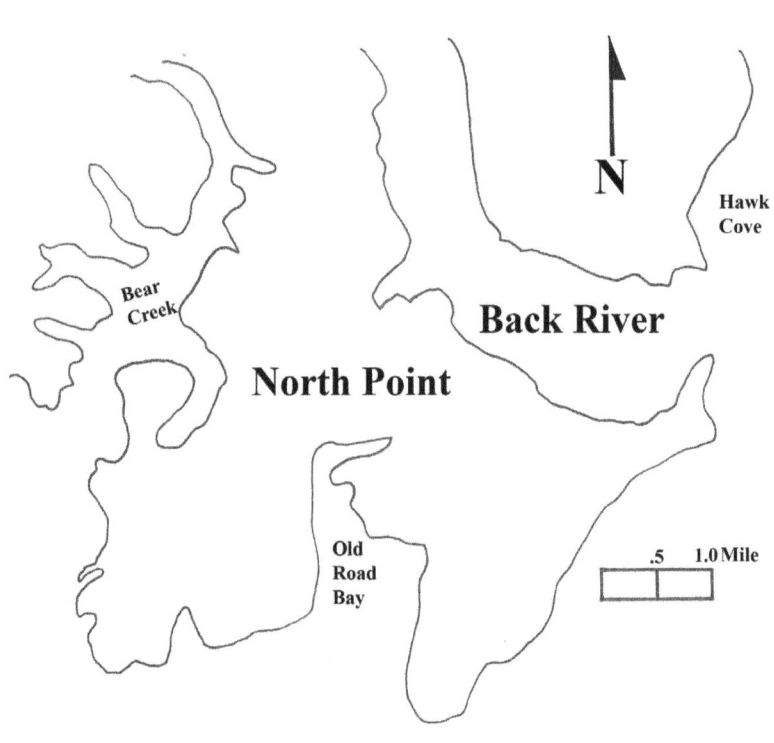

Map 57

Game Notes

The Battle of North Point lends itself to be reconstructed on the tabletop in a number of actions, including the meeting skirmish where Gen. Ross was mortally wounded, the main British advance on Baltimore at North Point, and the charge of the dragoons on the retreating British. An article in *Wargames Illustrated* 260 provides for a fourth scenario where a company of the Sharp Shooters (referred to as Baltimore Jagers in the article) skirmish with the light company of the 44th over control of a log cabin.

Scenario #1 – The Death of Gen. Ross

On the morning of the 12th, Gen. Ross and Adm. Cockburn halted at the Gorsuch farm and made their hosts serve them breakfast at a table in the front yard, while the rank and file helped themselves to the family's cattle, pigs and poultry. An American picket observed the actions of the British and reported what they saw to Gen. Stricker. Stricker assembled a strike force under Maj. Heath and sent them forward to upset the British late breakfast.

At this point, the officers had finished their breakfast and began to march toward the American defenses ahead of them. A mile down the road from the farm, the two forces blundered into each other. Mutually surprised, the two forces began shooting at each other. Ross, interested to see what the commotion was about, rode forward to observe. Witnessing the accurate American fire, he turned his horse around to bring up more men. At that point, a rifleman's ball did its job and Ross has hit. With the wounding of the British commander, the engagement reached a pause, and Heath was able to extract his troops back to the American defensive line.

Scenario Order of Battle
British, *580 men*
Commanding Officer: Maj. Gen. Robert Ross
Rear Admiral George Cockburn
Colonial or Royal Marines, 60 men
85th Foot, light company, 60 men
4th Foot, light company, 60 men

Reinforcements
85th Foot, 400 men

US, *222 men*
Commanding Officer: Maj. Richard Heath

Fifth Maryland Militia, two companies, *120 men*
First Baltimore Sharp Shooters, riflemen, 70 men
4-pdr. cannon, *12 men*
US Dragoons, small detail, *20 men*

Map 58

Recreating this true meeting engagement is rather simple. A single road runs across the length of the table. The area is farmland, with scattered walls and small clumps of trees. The British forces, entering on the eastern edge of the table, are moving on the road, with the British light companies deployed on the flanks. The Americans enter on the western table edge and deploy the riflemen on the flanks, while the remaining troops are on the road itself. The British reinforcement appear on the table two turns after the British officers are shot at.

Once the forces being firing at each other, Ross, Cockburn and other British officers ride up to the firing to assess the situation for themselves. If the group of officers is fired upon by the Americans, a die roll is required to see which of the privileged class is cut down. On a roll of a 1, 2, or 3, of the following officers is hit:

1 - Ross
2 - Cockburn
3 - Col. Brooke
4 - aide de camp Capt. Duncan MacDougall
5 or 6 – staff officer

133. First Baltimore Sharpshooters, courtesy of Richard Watts.

Being hit by a ball does not automatically translate into the figures being removed from the field. A roll of a second D6 will determine the fate of the casualty

1 - Head Shot. The troops witness the dastardly execution of their beloved commander and no longer need to test for morale for the rest of this game.

2 - Body shot. The troops notice the officer has been killed outright but are not affected.

3 - Body shot. The officer is carried to the rear and dies later. British troops cannot fire or advance for the next turn.

4 – Wounded officer. The officer is carried to the rear but returns to duty. For two turns the officer cannot give any orders for the troops under his command. The troops can only continue to fire and defend themselves if charged but cannot advance or retreat.

5 - Light wound. Command range for the officer is reduced by half for the rest of the game.

6 - Ball goes straight through the officer's hat. Carry on as usual.

Ross commands the entire army and can order any unit to act. Cockburn can only order the Marines and Brooke can only command the 85th.

Scenario Victory Conditions

The British need to drive the Americans from the field by the end of the game. The Americans gain a victory point for every two British rank and file casualties they cause, three points they kill Ross, and five points if they kill Cockburn, the most hated man in America at the time. The game plays for twelve turns.

American major victory - The Americans hold the field AND gain 20 Victory Points
American minor victory - The Americans hold the field OR gain 20 Victory Points
Draw – The Americans leave the field but fail to obtain 20 Victory Points
British minor victory - Drive the Americans from the field
British major victory - Drive the Americans from the field without Gen. Ross being killed

Scenario #2 – The Assault on North Point

The British assault started well, but a British unit got pulled from the battle and followed up on chasing a fleeing American unit. Additionally, accurate American rifle fire caused heavy officer casualties, which caused command and control issues among the British. The American troops were able to stand against the British attacks but had difficulties in moving across the battlefield and deploying to refuse their flank.

No special rules are needed to recreate this British attack. The American general recognized the limitations of his troops and used them to the best of their abilities. The absence of Ross was felt by the British, as his replacement Col. Brooke was not as experienced or aggressive.

Scenario Order of Battle

British 4,419 men
Commanding Officer: Maj. Gen. Robert Ross

1st (Light) Brigade
Maj. Timothy Jones
85th Regiment, 850 men
4th Regiment, light company, 60 men
21st Regiment, light company, 90 men
44th Regiment, light company, 60 men

2nd (Right) Brigade
Lt. Col. Thomas Mullins
4th Regiment, 650 men
44th Regiment, 600 men

3rd (Left) Brigade
Lt. Col. William Patterson
21st Regiment, 900 men
Royal Marines, 2nd Battalion, 725 men
Royal Marines, Provisional Battalion, 107 men

Naval Landing Force, Rear Admiral George Cockburn
Naval Brigade, 600 seamen
Colonial Marines, 300 men

Royal Artillery, two 9-pdrs, one 5-inch howitzer
Royal Rocket Artillery
Royal Sappers and Miners, 4th Battalion, 2nd Co.

US, 1,400 men
Commanding Officer: Brig. Gen. John Stricker

1st American Line - 1,500 men
Twenty-Seventh Maryland Militia, 500 men
Fifth Maryland Militia, 550 men
Baltimore Union Artillery, four 6-pdrs.
First Maryland Militia Artillery, *two 4-pdrs.*
Composite Maryland Militia Riflemen, *350 men*

2nd American Line - 1,150 men
Thirty-Ninth Maryland Militia, 450 men
Fifty-First Maryland Militia, 700 men

3rd American Line - 700 men

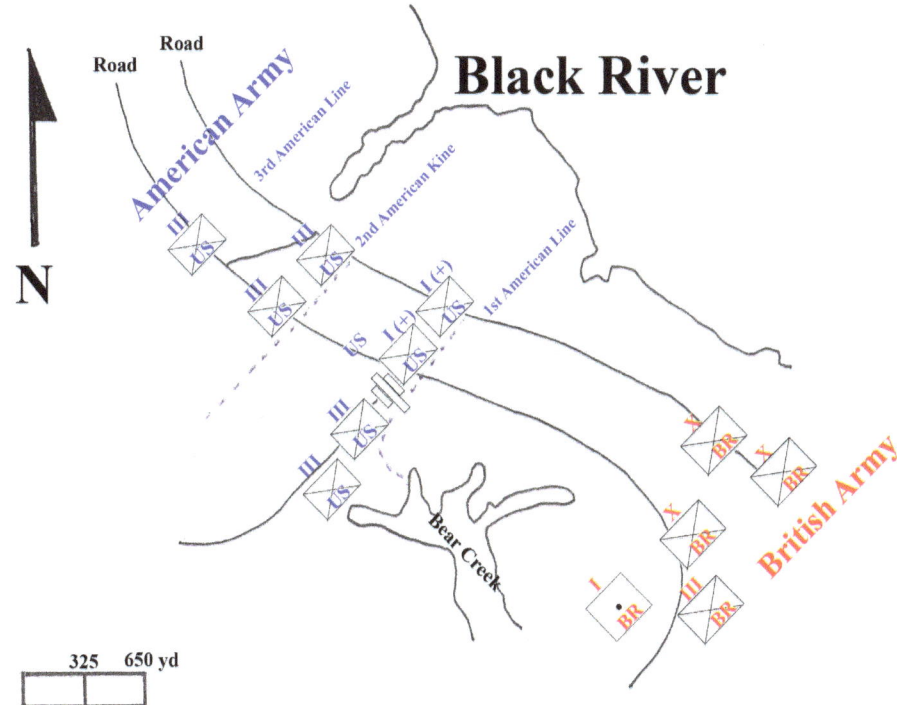

Map 59

134. Fifth Maryland, courtesy of Richard Watts.

Sixth Maryland Militia, 700 men

Scenario Victory Conditions
The British need to either destroy the American Army or drive them from the field by the end of the game. If the British fail to do either of these, the Americans win. The game goes for twelve turns.

Scenario #3 – US Dragoons attack the British Rear Guard
British troops, disappointed that they were unable to pillage Baltimore and acquire more spoils of war, began to march back to their boats. The Americans were uncertain of the actual British intentions and sent troops to follow. The Americans recognized that the British remained a coherent combat force, but for morale purposes, a show of military activity could be useful, especially for a general going into politics.

Rather than risk an all-out engagement, where the Americans could easily snatch defeat from the jaws of victory, the US forces focused on picking up British stragglers. When reaching the general area where Gen. Ross was killed, a thirty-man squadron of American dragoons caught up to the rear guard of the 21st Foot.

Here the dragoons charged through the British line. The British quickly unlimbered two cannons but the dragoons were able to charge through the British line once again. A third dragoon charge resulted in the capture of six British prisoners. Gen. Brooke halted his march and prepared for battle, but noting the dragoons were not supported by infantry, resumed his march to the British boats.

Mounted cavalry actions in the War of 1812 were uncommon. In the West, US troops under Harrison fought as traditional dragoons – riding their horses into combat but dismounting to fight. Along the Niagara Peninsula, both the British and Americans had cavalry, but they were used primarily for scouting and communication. At Sackett's Harbor, American dragoons fought on foot, as they did not have enough horses to equip the riders. At North Point, the US dragoons charged the enemy, sabers flashing in the sunlight.

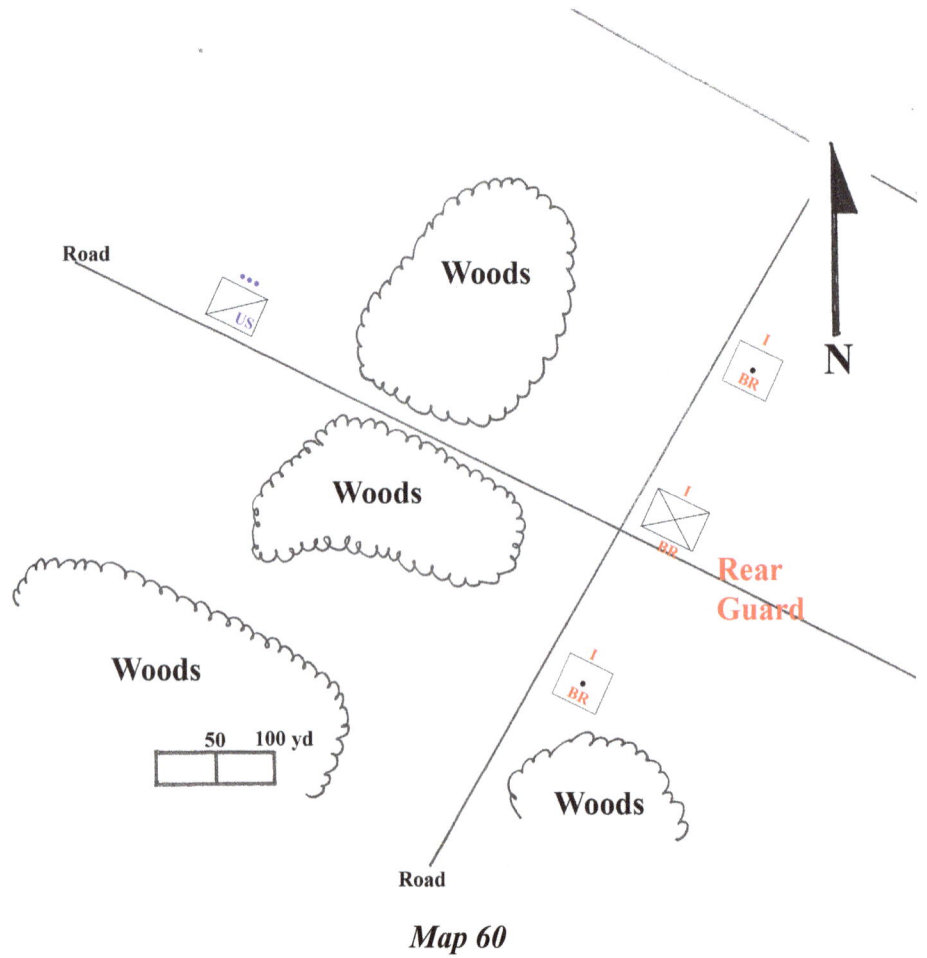

Map 60

135. 15mm British Infantry, courtesy of Dave Abraham.

Scenario Order of Battle
British
Commanding Officer: Unknown Lt.
21 Foot, *40 men*
two 6-pdr guns, *20 crew*

US
Commanding Officer: Capt. James Burd
US Light Dragoons, *30 men*

Scenario Victory Conditions
Both sides try to capture prisoners and escort them to their lines. Once a figure loses a melee, they are considered to be captured. Once captured, the victor must escort the prisoner to their lines. If the victor is taking one prisoner, the prisoner has no chance of escape unless the victor is killed.

One figure can guard up to three prisoners, but for every additional prisoner, there is an increased chance of attempted escape. Any time a prisoner crosses as obstacle, such a fence, road, or stream, a D6 is rolled for each man being escorted. On a 6, a prisoner will try to flee toward the nearest visible friendly comrade. If no friendly comrades are within a line of sight, the prisoner will move directly toward their edge of the table.

The side with the most prisoners wins the scenario. The game begins with the British marching on the road in column, with their rear to the enemy. The guns are unloaded and are being pulled; they will need a turn to deploy. The game plays for ten turns.

Further Reading
Elting, John R., *Amateurs, To Arms!* (Da Capo Press, Inc. NY. 1995)

George, Christopher T., *Terror on the Chesapeake: The War of 1812 on the Bay.* (White Mane Books. Shippensburg, PA. 2000)

Sheads, Scott S., *The Chesapeake Campaigns 1813-1815.* (Osprey Publishing. Oxford. 2014)

136. Royal Rocket Artillery, courtesy of Warlord Games.

Kirby's Windmill, 31 October 1814

Overview
Large Skirmish - less than 1,000 participants

Mounted US dragoon cavalry charge drives off a British raid.

Introduction
After the British failed to capture Baltimore in September 1814, most of the Royal Navy left the Chesapeake area. Cockburn sailed south while Cochrane sailed to Halifax in preparation for the assault on the Americans in the Gulf of Mexico. A number of British boats, however, remained in the area under the command of Capt. Robert Barrie. The remaining Crown forces renewed their attacks on the small settlements along the Chesapeake.

On 27 October, a force of Royal Marines and sailors landed at Tracy's Landing with the goal of destroying tobacco warehouses. A single US cannon fired upon the British landing barges and as the boats approached the shoreline, the handful of militia ran away. The British set fire to a number of buildings, including a tobacco warehouse, and occupied the town. In response to this raid, Virginia Governor James Barbour sent the First Brigade under Brig. Gen. William Madison across the Potomac to stop the British.

The Battle
On 31 October, a force of Royal Marines and sailors marched on Kirby's Windmill with the intent of capturing supplies and burning the structure. Along the way, they are charged by the US dragoons. A number of the British attempt to surrender but someone called for the dragoons to retreat. At this point, the horsemen obeyed, which allowed the British to rally behind a fence. A second cavalry charge was met with galling fire.

At this point, Brig. Gen. William Madison's forces appeared on the battlefield and the British speed up their withdrawal. Pursued by the dragoons, the British were able to board their boats and return to their ships.

Order of Battle
British, 300 men
Commanding Officer: Capt. Robert Barrie
Royal Marines and Royal Navy, 300 men

US, 1,000 men
Commanding Officer: Capt. James A. Burd
First Regiment US Light Dragoons, *120 men*

Brig. Gen. William Madison
First Virginia Brigade, 800 men
Battery of field guns, *four 6-pdrs., 80 men*

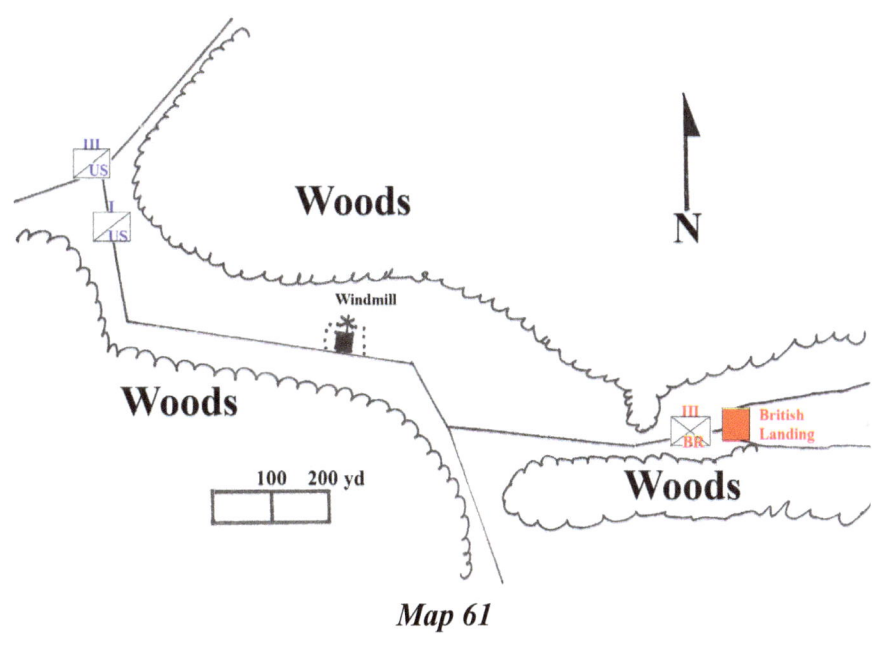

Map 61

Game Notes
With only a few passing references to the engagement, the wargamer can extrapolate from the known facts and create a decent small action tabletop recreation. Though cavalry did not play a significant role in the War of

137. Royal Navy Landing Party, courtesy of Wargames Foundry.

1812, at Kirby's Mill the US Dragoons drove back the British. One can also argue that the presence of Madison's 800 men strong First Virginia Brigade could have deterred the British from pressing their advance but given the low opinion the British had of the American militia, their presence may not have played a critical role. Indeed, the British were already retreating when they were first charged by the dragoons.

The game begins with the Royal Marines advancing along a road toward the mill. The British need to reach the mill and spend a turn looting/setting ablaze the structure. With their booty in hand, the British then need to move back to their landing site. It should take the Marines three turns to reach the Mill, one to loot/burn and three more turns to get back.

For the Americans, the dragoons appear on the road at the other end of the table at the start of the game. Madison's First Virginia Brigade appears only at the start of turn six on the other side of the table. The mill is in the center of the table. The horsemen move faster and can reach the mill in two moves. The area is rolling fields with few clumps of trees and a few scattered fences.

The game plays for ten turns.

Victory Conditions

The British need to get their supplies, set the mill ablaze, and get back to their boats. The Americans must drive the British out. Theoretically, both sides can claim victory but to gain a major victory, one must stop the enemy from achieving their victory conditions.

Further Reading

Poulos, Anastasia and John E. Kille, The British Were Here!: The War of 1812 in South Anne Arundel County. *Anne Arundel County History Notes*. Vol. XLV, No. 3. 1-10. 2014

Sheads, Scott S., *The Chesapeake Campaigns 1813-1815*. (Osprey Publishing. Oxford. 2014)

138. Virginia militia, authors photo.

Point Peter, 13 January 1815

Overview

Delaying action - less than 2,000 participants

A small group of US Infantry and Riflemen try to hold off a large British force.

Introduction

At the beginning of the 19th century, with Florida under Spanish control, the town of St. Marys, Georgia was the southernmost US settlement on the eastern seaboard. To protect the American border, an earthen and timber battery was constructed on Point Peter (also called Point Petre), where Peter Creek empties into the St. Marys River.

On the eve of the War of 1812, St. Marys was the third most protected US harbor and eleven gunboats were stationed to protect US interests. In 1813, Maj. Gen. Thomas Pinckney, commander of the 6th Military District of the United States, established his headquarters at Point Peter, but later moved it to Camp Hope. In November 1813, two batteries, one of which was covered by a stockade, were noted in the vicinity of St. Marys.

139. View toward Point Peter and St Mary's from Cumberland Island, courtesy of Cumberland Island National Seashore, National Park Service.

By January 1815, the US government made plans to fortify its southern most outpost. However, British troops remained in the area and continued to conduct military operations. To draw American troops from possibly reinforcing the British advance on New Orleans, on 10 January 1815 Admiral Sir George Cockburn landed 1,500 men on Cumberland Island off the Georgia coast. Though the British were defeated at New Orleans five days earlier, on 13 January Cockburn began his advance.

The Battle

Admiral Cockburn's forces first bombarded and then landed near the fort at Point Peter. The ranking US commander in the area, Capt. Abram A. Massias, recognized the threat and the disparity of forces. Commanding both his company of US Riflemen and a company of the Forty-Second US Infantry Regiment, Capt. Massias organized a two-level defense which maximizing the advantages of the local terrain. A thirty-six man detachment was positioned at the battery, which was to fire at the enemy and spike the guns before being overrun. The remaining troops were placed on both sides of small gorge that the British troops needed to traverse in their advance.

As the British troops advanced in column, the Americans opened fired. The British troops ignored their losses and pushed forward. Massias then withdrew his forces into the surrounding woods where they continue to fire at the British. The Marines pursued the Americans into the woods and the Americans withdrew in the face of superior numbers. Having captured the American position and dispersed the enemy, the British then resumed their march to St. Marys.

Arriving at St. Mary's, the British seized two American gunboats and twelve merchantmen, including the *Countess of Harcourt,* a former British East India ship which was recently captured by an American privateer. The

British troops proceeded to loot the jewelry store, steal fine china from the residence, and burn the public works, including blockhouses and barracks. After a week's occupation, the British withdrew to Cumberland Island, taking 1,500 former enslaved people with them.

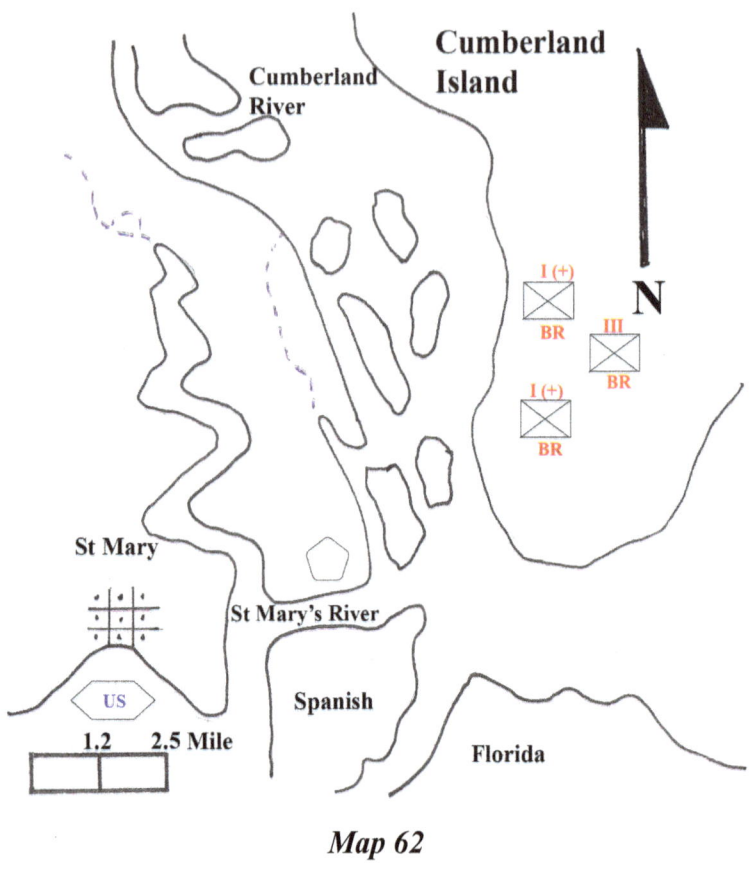

Map 62

Order of Battle

British - *1,200 men*
Commanding Officer: Admiral George Cockburn
Royal Marines, 3 battalions, each of 4 companies, 840 men
Royal Marines, ship's detachments, 2 companies, 120 men
2nd West India Regiment, 2 companies, 190 men

US - 126 men
Commanding Officer: Capt. Abram A. Massias
Forty-Second US Infantry Regiment, 1 company, 36 men
US Rifles, 1 company, *80 men*

Game Notes

At first, this engagement does not appear to make for a balanced tabletop encounter. The American forces are outnumbered ten to one and the British are led by a legendary Admiral. Though there is no evidence that Cockburn was personally involved in the engagement, having him physically represented on the tabletop adds a bit of dramatic flair for the game, and an asset that the British player can lose.

Capt. Massias correctly observed that he cannot hope to stop the British advance and the most that he can do is to make a good showing. 19th century honor demanded that he made a demonstration before the enemy before retiring from the field. Both of these actions Massias accomplished successfully. Given the conditions in which he operated, Massias accomplished his mission and thus "won".

The British also "won" their short-term tactical objectives – they destroyed the US military assets in the area and captured twelve merchant ships. By removing approximately 1,500 former slaves from the US, the British also gained the moral high ground and to a small degree, undermined the US economic structure. The British could have also pursued the US Riflemen further, but as it was not in their battleplan, they let them escape.

In replicating this engagement on the tabletop, we need to move beyond the traditional established victory conditions and look at points of honor. The American player cannot run away from the overwhelming enemy but must make a demonstration. The more casualties the US player causes, the more victory points they achieve. In our tabletop recreation, the American player gains a point for every British casualty suffered. As the American player has chosen to make a demonstration against the enemy, they receive 10 victory points at the start of the game.

Conversely, the British player needs to destroy the American military presence in the area and casualties be damned. However, for every US unit and public work (barracks, battery, and blockhouse) destroyed or pushed off the table, the British player gains a point. The boats in St. Marys were doomed to be captured, as the British squadron was blocking access to the sea, but the American player can send some of this troops to move the two gunboats further upstream to prevent their destruction. To move the gunboats, however, US troops need to be sent to accomplish this task and if so deployed, they do not return for the rest of the game. The Americans gain 5 victory points by saving the boats. The British have no honor points at the start of the engagement.

Having Admiral Cockburn on the tabletop adds an interested twist to the encounter. If the Admiral commands from the front, the British player receives two victory points every turn. For every unit ahead of Cockburn, subtract one victory point - thus if two British units are in front of the admiral, then the British player earns no victory points for that turn

The riflemen were chosen by their ability to shoot. Though the specific targeting of officers was considered not very sporting, it was commonly done, and the riflemen excelled at shooting officers. Every time a rifleman has the opportunity to shoot at the admiral and does so, the US player loses a victory point. However, if Cockburn is killed, the American gains 10 victory points. If a rifleman (or a US Infantryman for that matter) injures or kills the admiral, the game slips into an alternative reality.

The death or even injury of Admiral Cockburn would greatly upset the British assault. Each British unit within 36 inches

140. US Rifle Regiment, courtesy of Roger Chrysler.

of the admiral will need to test their moral. On a roll of a 4-6 on a D6, the unit in question will charge the nearest US troops, casualties be damned. On a roll of a 2 or 3, the British troops are simply stunned and sit still for a turn. If a British unit rolls a 1, then they run away from the enemy. At the start of the following turn, the routing troops may try to reform (a 4, 5, or 6 on a D6 is needed). If successful, the troops need to spend the rest of this turn forming ranks and cannot do anything else; if they are unsuccessful in their die roll, they will move away from the enemy and can try again to reform on the following turn. If they fail for two turns to stop running, they are removed from the tabletop and the US player gains one honor point for every two figures routed.

In the early 19th century, looting and stealing from civilians, though considered to be distasteful, was an all-too-common occurrence. Both British and American troops, like the multitudes of generations before them, commonly used the opportunity presented on campaign to enrich themselves. Civilians caught in the middle, or perceived to be defending themselves, often suffered. Admiral Cockburn's troops had a history of not respecting civilian property and according to a number of American accounts, helped themselves to civilian goods at St. Marys.

For game purposes if, or more likely, when the British enter St Marys, each unit needs to test if it breaks ranks and begins to loot (on a die roll of 3, 4, 5, or 6 on a D6 the British unit begins to loot). For every two units looting, the British lose a victory point. In our game St. Marys was represented on the tabletop with a blockhouse, two barrack buildings, and eight stores and homes (this includes the jewelry shop). The game ends when the British set fire to the blockhouse and the barracks.

141. 2nd West Indian Regiment, courtesy of Tod Kershner.

Victory Conditions

The player with the most Honor Points wins the scenario. The game ends when the British set fire to the blockhouse AND the barracks.

Further Reading

De Quesada, Alejandro M., *A History of the Georgia Forts*. (The History Press. Charlestown, South Carolina. 2011)

Elliott, Daniel T., Point Peter and the St. Marys River Forts. LAMAR Institute Publication 62. Box Spring, Georgia. 2002)

Fredricksen, John D., *Green Coats and Glory: The United States Regiment of Riflemen, 1808-1821*. (Old Fort Niagara Association. Youngstown, New York. 2000)

Gulf Coast

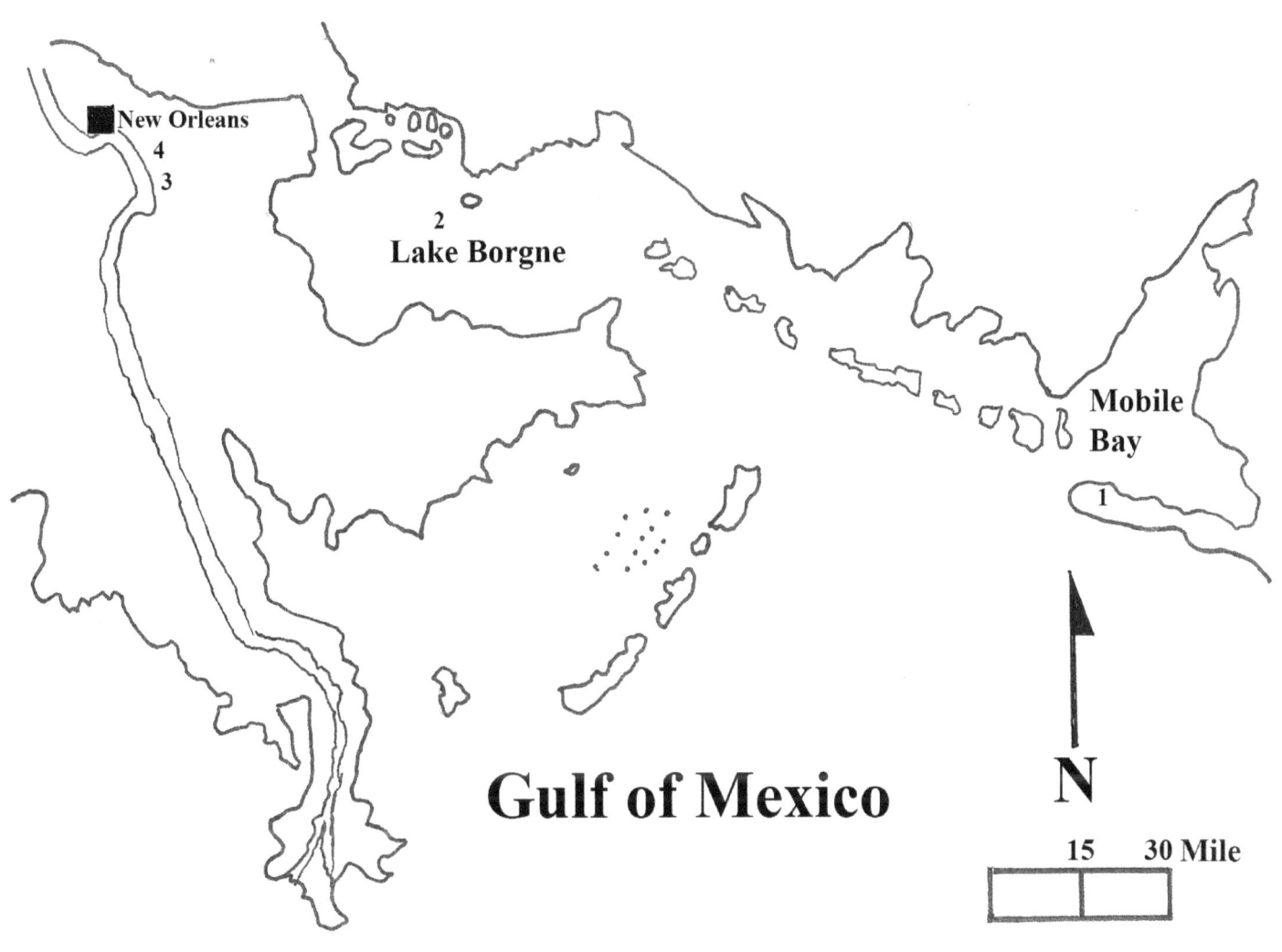

1. Ft Bowyer   2. Lake Borgne   3. Villere Plantation   4. New Orleans

Map 63

Ft. Bowyer, September 1814

Overview
Small Skirmish - less than 500 participants

Royal Marines and Native Americans try to capture an American fort protecting Mobile, AL.

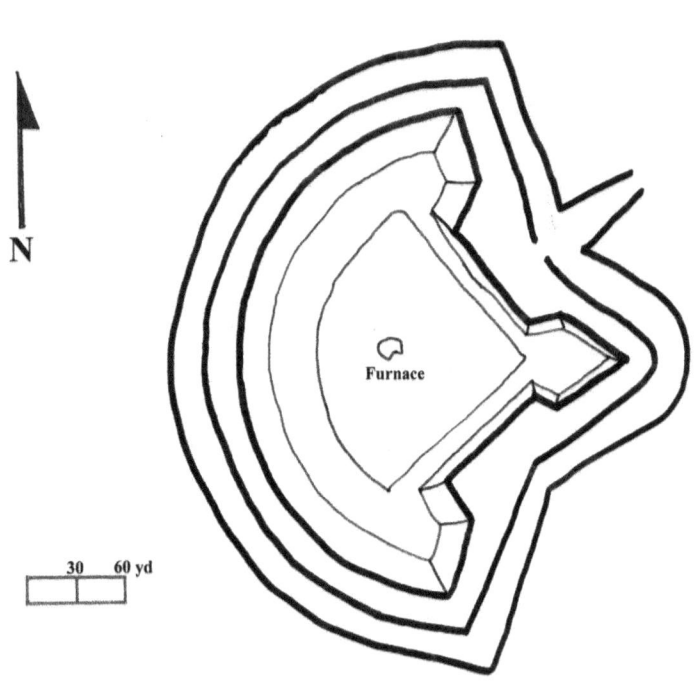

142. Plain of Ft Bowyer, drawn by Christian Wright.

Introduction
　　The War of 1812 took place against a backdrop of American expansionist tendencies that focused not only on territories controlled by Britain, but those governed by Native Americans and the Spanish Empire. In 1810, President Madison coordinated an insurrection in Spanish West Florida and as Spain was dealing with Napoleon's occupation, the land came under US jurisdiction. In early April 1813, Maj. Gen. James Wilkinson arrived at Mobile and informed the Spanish he was there to relieve the half-starved garrison. Once the Spanish were evacuated to Pensacola, Wilkinson ordered the construction of a semi-circular fort to guard the entrance to Mobile Bay. By June 1813, Col. John Bowyer had completed the sand and log fortification.
　　During the 1813 Creek War, the Americans defeated the Native Americans and many refugees fled to the still Spanish controlled territories of East Florida. Charles Cameron, the British Royal Governor of the Bahamas (Nassau), called for the British to send military personnel into the region for the express purpose of arming and training potential allies against the American expansion. In April 1814, Capt. Hugh Pigot of the Royal Navy began to arm the Creeks from a base on the Apalachicola River and Pensacola. In August, Maj. Edward Nicholls arrived with a battalion of Royal Marines to take command of the operation. As British Admiral Cochrane was completing his operations in the Chesapeake, the forces of the Crown were planning on larger operations in the Gulf of Mexico. On 3 September, the British sought to enlist the services of the pirate Jean Lafitte to their cause.
　　Maj. Gen. Andrew Jackson, recently appointed to command the 7th Military District, focused his attentions on capturing Pensacola. On 22 August, Jackson moved the district headquarters to Mobile, with the plan of using the city as his base of operations. Capt. Sir William Percy, RN, and his four sloops-of-war, also headed for Mobile. Before they could get to the city, however, they would need to capture Ft. Bowyer.

The Battle
　　The British planned to attack the fort from both the land and the sea. Percy landed a contingent of Royal Marines, Native Americans, and a howitzer cannon behind the fort and then began to bring his boats to fire upon the fort. Navigating the maze of sand bars in front of the fort, Percy was able to bring the *HMS Hermes* to within musket shot range. The *HMS Sophie* was close behind. In the artillery duel that followed, the *Hermes* was aground and on fire. With the *Sophie* damaged, Percy picked up the landing party and retreated.

Epilogue

After the American victory at New Orleans, the British focused their attention on the capture of Mobile and once again attacked Ft. Bowyer. Though Jackson had placed additional men inside the fort and vowed that it would never be taken by even 10,000 men, the coastal fortification lacked underground powder magazines and was vulnerable to regular siege operations. The British successfully landed a brigade of troops and after digging entrenchments, mounted their batteries. The Americans surrendered the next day. The day after, on 13 February, a British dispatch ship arrived and brought news that the war was over.

Order of Battle

British
On Land - 202 men
Col. Edward Nicolls, RN
Royal Marines, 60 men
Native Americans, *130 men*
Howitzer, 1 gun and 12 crew

At Sea
HMS *Hermes*, sloop-of-war, 22 guns
HMS *Sophie*, brig-sloop, 18 guns
HMS *Carron*, sloop-of-war, 20 guns
HMS *Childers*, brig-sloop, 18 guns

US - *200 men*
Commanding Officer: Maj. William Lawrence
Second U.S. Infantry Regiment, 160 men
12 antiquated heavy guns, *40 men*

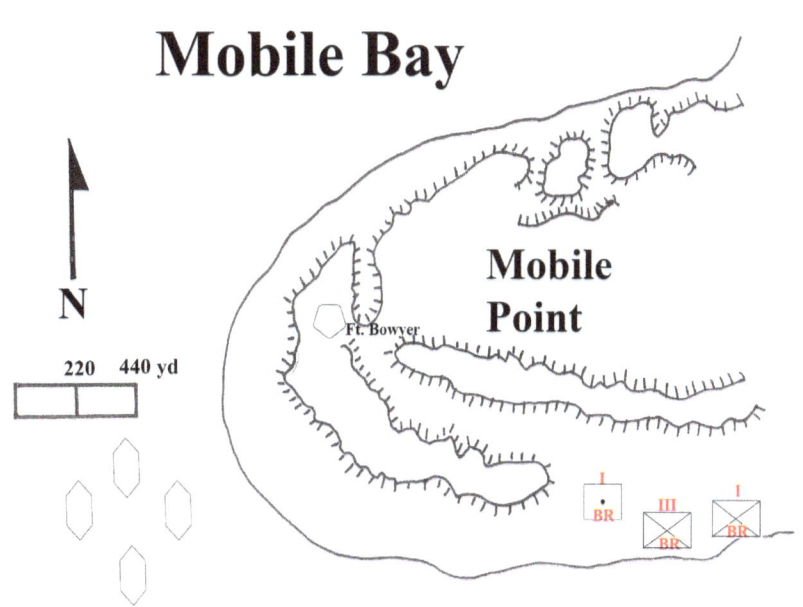

Map 64

Game Notes

Gaming the second British attack does not make for much of a game. Pitting 375 men and a few cannons to defend a poorly protected area against 14,000 professional troops in terms of game play, is both impractical and silly. In the first battle, which is presented here, both sides have a somewhat equal potential of winning.

Historically, the British held off launching their land-based force and never attacked. In this tabletop recreation, however, the Royal Navy continues to bombard the fort while the Royal Marines and Native Americans attack from the land. To breach the back of the fort, the British have a howitzer that should be able to break down the wooden gate protecting the rear of the fortification. All the Native Americans are armed with muskets and various small arms (hatchets and war clubs). Though historically the Royal Marines and Native Americans were positioned more to the south of the fort, in our tabletop recreation, they place them more to the north, which puts them closer to the fort's wooden gate. In our game, the landing force is 26 inches away from the gate.

The British can scale the sand and timber walls – go over the top – or try and breach the wooded gate with their cannon. In

143. Native Americans courtesy of Mike Vasile.

game terms, if the gun is within 4 inches of the gate, anything but a 1 is needed on a D6 to blow open the gate. For every two additional inches the gun is placed further from the gate, an additional pip is needed. Thus, at 6 inches away, the die roll needs to be anything but a 1 or a 2, and at 8 inches, anything but a 1, 2, or 3. At 14 to 24 inches, the gun needs to roll a 6 to damage the door and a second shot without any modifiers, to punch it open. If the Royal Marines and Native Americans chose to go over the top, it will cost the British one additional turn to reach the top of the defensive wall and a second turn to go down into the fort itself.

The actual number of cannon available to the Americans is disputed, but at least one writer noted that they were antiquated and possibly some or perhaps even all of them were unsafe. If your game rules allow for exploding guns, it would be appropriate to use them in this scenario. If not, a simple work around would be to reroll any 1s that appear on a die roll. On this second die roll, if a 1 appears, the gun and crew are considered to be destroyed, if a 2 is rolled, then the gun can be put back into service next turn but will require a new crew to be pulled from the fort's defenders.

The game begins with the Royal Navy engaged in an artillery duel with the fort. Ten men are needed to operate each of the 12 cannons. For every British naval hit on the fort, the Americans may suffer a crew loss and then would need to either redeploy their infantry to crew the guns or reduce their fire by one gun.

Victory Conditions

Whoever controls the fort at the end of the game wins. The game plays for twelve turns.

Further Reading

Chartrand, Rene, *Forts of the War of 1812*. (Osprey Publishing. Oxford. 2012)

Elting, John R., *Amateurs, To Arms!: A Military History of the War of 1812*. (Da Capo Press. New York. 1995)

Lossing, Benson J., *Pictorial Field-Book of the War of 1812*. (Harper& Brothers Publishing. New York. 1869)

Stoltz, Joseph F., *The Gulf Theater 1813-1815*. U. S. Army Campaigns of the War of 1812. (Center for Military History, Washington. 2014)

144. US Infantry wearing white linen or cotton jackets, author's photo.

Battle of Lake Borgne, 14 December 1814

Overview
Naval Engagement

Forty-two British longboats overrun five American gunboats and gain control of Lake Borgne.

Introduction
　　Learning that the Americans were poised to resist a British landing at Mobile, Vice Admiral Sir Alexander Cochrane decided to send his forces to Lake Borgne. Upon his arrival on 13 December, Cochrane spotted the American flotilla returning to their base. The British boats began to give chase but got stuck on sandbars. The British were able to free themselves only to get stuck again as the tide retreated.
　　The next morning the British launched 42 longboats, with each boat mounting a single carronade on its bow. After rowing toward the American squadron, Capt. Lockyer ordered the men to halt and eat breakfast. After finishing their meal, the British rowed within cannon range and began firing.

The Battle
　　The Americans had arranged their five gunboats in a battleline between Point Claire and Malheureux Island. A sixth craft, a dispatch boat was placed in reserve. The British divided their craft into three waves and the two sides began firing at 1100. Though the Americans were able to sink a number of barges and broke up the initial assault, the British craft pushed forward and boarded the American vessels. Fifteen craft attacked the center of the American line. Upon the capture of *Number 156*, the British were able to turn the gunboat's cannons on the remaining American craft and began firing. In the course of the action, both commanders were wounded. In the end, the Americans were unable to resist the superior numbers and the British carried the day. As the battle was lost, the Americans set fire to the lightly armed dispatch boat to prevent its use by the enemy. With the American naval presence on the lake neutralized, the British were free to advance on New Orleans.

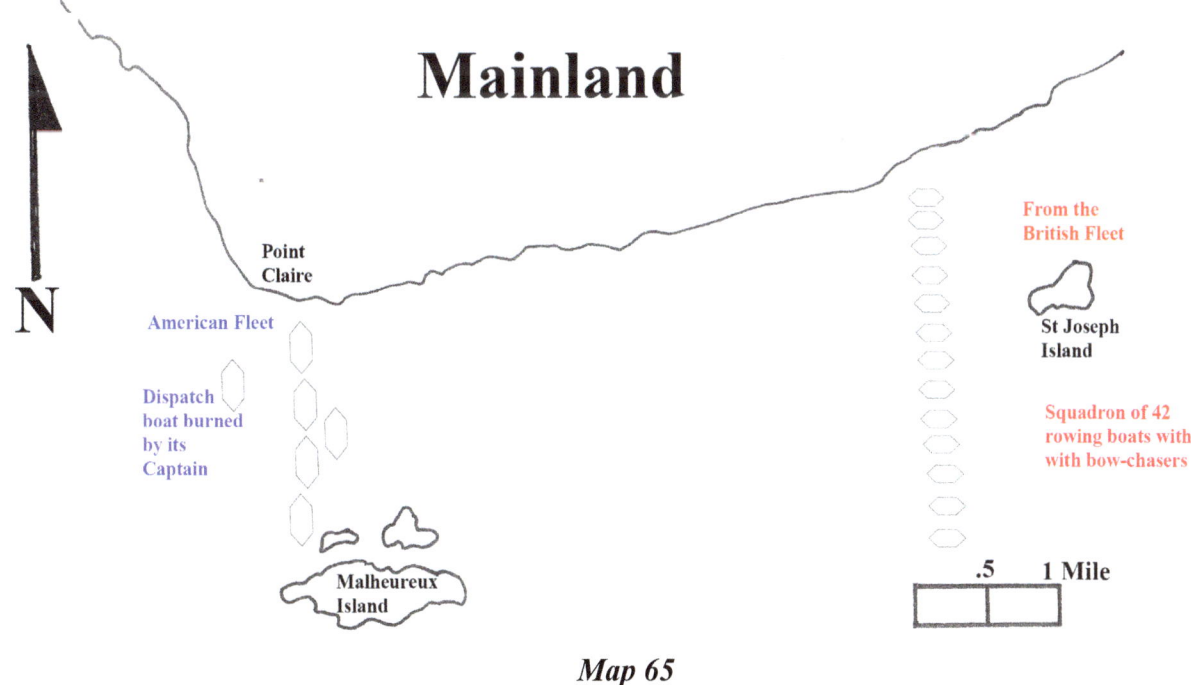

Map 65

145. US Gunboats of the types used during the War of 1812, 19th century print, Naval History and Heritage Command.

Order of Battle
British 1,200 men
Commanding Officer: Capt. Lockyer
42 longboats, each with one carronade

US 182 men
Commanding Officer: Lt. Thomas Ap Catesby Jones
5 Jefferson gunboats, *Number 156, Number 163, Number 162, Number 5,* and *Number 23.*
    Number 156 was armed with one long 24 pdr., four 12 pdr. carronades, and four swivel
        guns.
USS Tickler, dispatch boat

Game Notes
    In essence, this is a game of a last stand fought on water. The Americans have five larger, better armed boats, but because of the low tide, they are unable to maneuver their craft. The British have the numbers, with some 1,200 sailors and Royal Marines ready to overrun the Americans. The Americans can try to reduce the number of attackers, but eventually some boats will reach the defenders. Once the Royal Marines begin boarding the gunboats, the Americans can no longer man their cannon and must fight hand to hand.
    To deter the British from closing, the Americans need some luck – both with wind and their cannon. While the British longboats are powered by oars and not sail, the wind has a significant impact on both speed and direction. If the wind is blowing against the longboats, they lose six inches of movement for every level of wind above breeze. If the wind is blowing with the landing craft, they approach the Americans much faster and gain six inches of movement for every level of wind above breeze. If the wind is blowing to the side, the boats drift six inches to the east or the west of the landing site. To keep the game moving, one die roll is applied to the entire British flotilla.
    To simulate wind speed and direction, various game mechanism can be employed. At the start of every turn, a D6 is rolled to determine the strength of the wind. Based on the die roll, wind patterns are established for the turn in question. On a roll of a 1 or a 2, the wind dies down, on a 3 or 4, the wind stays the same, and

on a 5 or 6, the wind increases. As in the other scenarios in this book, four levels of wind are noted – no wind, slight breeze, slight wind, and strong wind. The game begins with no wind. The British boats start off six turns away from the Americans.

To determine the wind direction, 2D6s are rolled and the results are applied at the start of the turn: a 2-3, wind turns 90 degrees counter-clockwise, on a 4-5, they turn 45 degrees counter-clockwise, 6, 7, or 8, they continue to blow in the same direction, on a 9 or 10, they blow 45 degrees clockwise and on an 11 or 12, they blow 90 degrees clockwise.

Once the British are with 2 inches of an American boat, they can attempt to grapple and board. Most naval games include a mechanism for boarding and conducting ship to ship melees. Each American boat is considered to be four times as large as one British longboat but as each British boat carried a complement of Royal Marines, each British longboat has twice as many crewmen that can engage in a melee. Thus, if two British longboats board one American boat, the British will have four times as many melee points.

146. Gunboat Squadron, courtesy of Warlord Games.

Victory Conditions
The British have twelve turns to destroy the American flotilla.

Further Reading
Lossing, Benson J., *Pictorial Field-Book of the War of 1812.* (Harper& Brothers Publishing. New York. 1869)

Pickles, Tim, *New Orleans 1815: Andrew Jackson Crushes the British.* (Osprey Publishing. Oxford. 1993)

147. A swarm of gunboat, author's photo.

Villere Plantation, 23 December 1814

Overview
Battle - 5,000 participants

Jackson launches a three-pronged night attack on a British camp.

Introduction

148. Andrew Jackson by Ralph Eleaser circa 1817, National Portrait Gallery.

Following the large-scale British raids along the Chesapeake in the late summer of 1814, Admiral Cochrane returned to Halifax, which served as the British headquarters of the North American Station to prepare for an expedition against New Orleans. Cochrane then sailed to Jamaica, where he picked up fresh troops, including two West India regiments which were composed of formerly enslaved Africans.

New Orleans was the major port of the western part of the United States. For two years, the warehouses were packed with raw materials and their capture would provide Cochrane with a great deal of prize money. For the Crown, the capture of New Orleans would close Mississippi River traffic to the Americans. Reports made available to the British also highlighted that much of the local population, including both the French and Spanish settlers and the Native Americans were eager to throw off the yoke of the United States.

At the end of May 1814, Andrew Jackson was assigned command of the 7th Military District and spent most of his time mopping up his Creek campaign and planning the capture of Spanish held Florida. The fact that the US was already in a struggle with Great Britain did not concern the American General. Building on Maj. Gen. James Wilkinson's previous capture of Mobile, Jackson focused on expansion. Calling for additional militia troops from surrounding states, in early November Jackson moved and captured Pensacola from the Spanish. With western Florida under US control, Jackson disbursed parts of his force to garrison various coastal settlements. Only in early December did Jackson return to New Orleans.

The British had a few options on how to best advance on New Orleans – either try to force their way up the Mississippi River or over Lake Borgne, a shallow body of water that emptied into the Gulf. Admiral Cochrane chose Lake Borgne. After the British defeated the collection of small US naval boats on Lake Borgne, Cochrane identified a good landing place and put his men ashore. Moving inland, the British learned that a company of American militia were stationed at Villere plantation. The American troops, following "the ancient militia custom of sleeping soundly while on guard duty" were captured by Col. William Thornton, who commanded the British advanced guard. While humiliated, American Revolutionary War veteran Maj. Rene de Villere managed to escape from the British and warn Jackson.

Between the stories told by refugees, Maj. Villere's statement, and Maj. Latour's report, a former intelligence officer sent by Jackson to access the situation, Jackson put together an overtly complicated three prong plan to attack the British. Though Col. Thornton urged his superior Maj. Gen. John Keane to lead the British in an immediate attack on the Americans, Keane chose to wait for more British reinforcements to arrive. In the late afternoon, a small number of American dragoons probed the area and were driven off by the forces of the Crown. As the British settled in for the evening, the Americans moved in for a night attack.

Jackson's plan called for the American Infantry and cannon to move down along the main road, which ran parallel to the river. Coffee's mounted Tennessee riflemen would move to envelop the British right flank. Gunfire from the *Carolina* signaled the start of the American attack.

The Battle

Darkness fell early and by the early evening when the British observed a boat in the river drifting towards, they believed the craft was from their squadron bringing in supplies. Using the British campfires as visible targets, the USS *Carolina* opened fire. Shortly thereafter, Jackson's men began to move on the British camp.

At this point, Col. Thornton took command and began to set up the British defense. Rallying the 85th and the 95th, Thornton established a defensive line and deployed the 4th in reserve. The US regulars pushed the British line in, but Thornton counterattacked. Units on both sides broke, rallied, and rejoined the fray. At one point, Jackson was in the main battleline, rallying the Americans and pressing forward. Between the fog, night, and the clouds of black powder, visibility was non-existent.

Along the American left flank, Coffee's troops dismounted and attacked Thornton's right. Units broke apart, but the men continued to fight in small pockets. Using an old abandoned levee, Thornton was able to put together a defensive line. With the arrival of more British troops on the field, Coffee's men disengaged successfully from the fight. Jackson's men were also able to pull back from the British.

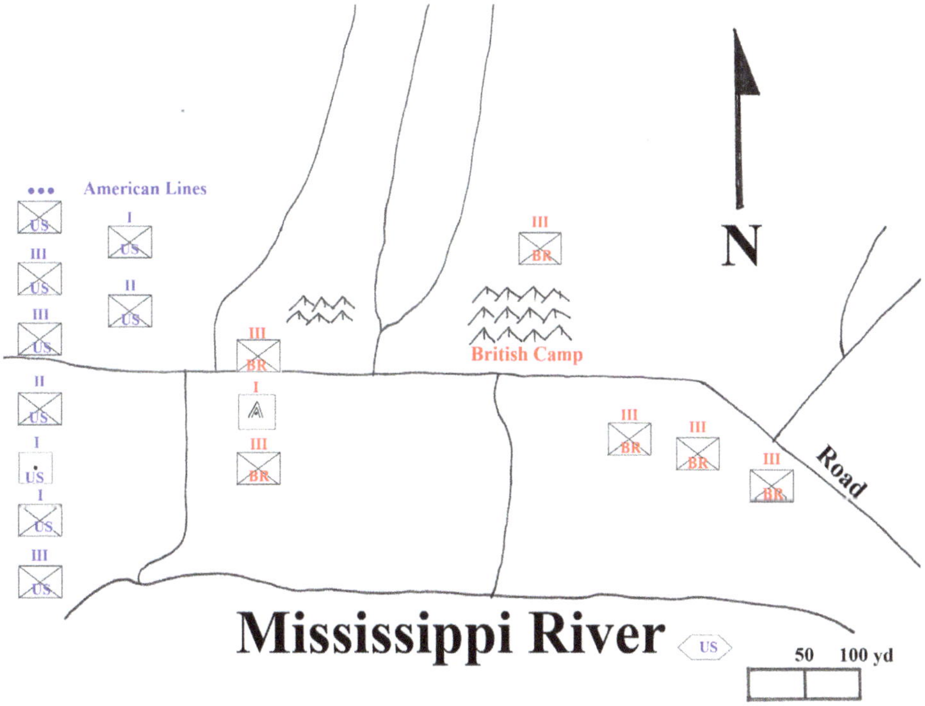

Order of Battle
British 3,400 men
Commanding Officer: Col. William Thornton

At the plantation
Col. William Thornton
85th Light Infantry, 650 men
95th Rifles, 500 men
Rocket Company, Congreve rockets, 80 men

In reserve
Maj. Gen. John Keane
4th Foot, 750 men

At the table edge farthest from the Americans
93rd Foot, only 200 men engaged in the battle
21st Foot, 900 men
44th Foot, 750 men

US 2,200 men
Commanding Officer: Maj. Gen. Andrew Jackson
Seventh US Infantry Regiment, 450 men
Forty-Fourth US Infantry Regiment, 400 men
US Artillery, two 6-pdrs.

Map 66

149. 95th Rifles, author's photo.

US Marines, 1 company, 60 men
Volunteer militia, 1 battalion, 400 men
Volunteer Free Men of Color Militia, 1 battalion, 200 men
Beale's Orleans Volunteer Rifles, 1 company, 40 men

Brig. Gen. John R. Coffee
Tennessee Mounted Riflemen, 9 companies, 600 men
Choctaw warriors, 18 men

USS Carolina, schooner, 14 guns

Game Notes

Under armchair analysis, Jackson's attack had all the hallmarks of a complete fiasco. It required coordination and timing between three separate forces. The plan somehow worked and had Jackson not scattered his US regular troops to guard far-flung settlements like Mobile, Baton Rouge, and Pensacola, their presence may have tipped the battle in favor of the Americans.

Individual British veteran troops performed well when faced with a night attack, but it was Thornton's leadership that held the line. Appearing everywhere when needed, Thornton marshalled the troops and held the line long enough for additional British units to be plugged into the battle.

To properly recreate the battle on the tabletop, the Americans side should be represented by two or three players. Though a single British player can control all of the Crown's forces, the disorganized nature of nighttime fighting is best represented when each British player controls no more than two or three units.

Fighting at night was fraught with issues. With the low light conditions, units arriving on the field could easily mistake an enemy for a friend or vice-versa. Even with a bright moon, targets were hard to see. Add the lingering smoke generated by weapons firing black powder and, in this case, some fog to the equation, and it's a wonder that anyone was actually hit by a projectile. Units clashed directly as both sided entered into melees with whatever weapons were close at hand.

Recreating such an event on the tabletop requires special consideration. The three conditions that need to be addressed are how troops react when they arrive on the tabletop, how the dark conditions impact a unit's ability to fire, and how the darkness effects melees. When a unit advances toward the enemy, there is a chance it will turn in the wrong direction and expose its flank. To replicate this on the tabletop, once a unit moves into a charge move distance from the enemy, it needs to roll a D6 and apply the following results.

1 – the unit turns 90 degrees to its left
2 – the unit turns 45 degrees to its left
3 or 4 – the unit faces directly ahead
5 - the unit turns 45 degrees to its right
6 - the unit turns 90 degrees to its right

Given the darkness and smoke that covered the battlefield, any unit that engages in firing may end up firing upon a friendly unit. To replicate this on the tabletop, any unit that fires needs to roll a D6 – on a roll of a 1 or a 2, the unit will end up shooting the closest unit from their side. All musket and rifle firing is limited to short range. Finally, melees fought in the darkness should only last one round. If no winner of the melee emerges from the first round of combat, both sides fall back a full movement rate facing the enemy.

The British reserves can move toward the Americans at the start of Turn 5. The old levee is considered to be hard cover. Once the US land forces start attacking, the *Carolina* stops firing. Both the 95th and the 85th are considered to be disordered at the start of the game. The game begins with the US land forces approaching the plantation.

Victory Conditions

After the battle, both sides claimed victory. The British claim rested on the fact that they continued to hold the battleground while the Americans claimed their attack delayed the timetable of the British assault, which allowed them to construct fieldworks at Chalmette Plantation.

On the tabletop, each side needs to drive at least half of the enemy off the table to win a brilliant victory. Failing to achieve this goal, the British or Americans can win a minor victory if they cause more casualties on the enemy than they receive. The game plays for ten turns.

Further Reading

Casey, Powell A., *Louisiana at the Battle of New Orleans*. (Eastern National Park and Monument Association. 1995)

Elting, John R., *Amateurs, To Arms!: A Military History of the War of 1812*. (Da Capo Press. New York. 1995)

Pickles, Tim, *New Orleans 1815: Andrew Jackson Crushes the British*. (Osprey Publishing. Oxford. 1993)

150. US Infantry, courtesy of Knuckleduster Miniatures.

New Orleans, 8 January 1815

Overview
Large Battle - 15,000 participants

British attempt to capture New Orleans.

Legacy
For Americans, the Battle of New Orleans is the best-known battle of the entire war in the US. Fought after the Treaty of Ghent was signed (but not yet ratified by Congress), the event has been enshrined for more than 200 years. Andrew Jackson was able to use the victory to win an election and become president. For much of the early 19th century, the victory itself was celebrated as a second 4th of July. In the 20th century, the battle was recognized in popular movies about Jackson. In 1958, the pirate Jean Lafitte got his own Paramount movie *The Buccaneer* and in 1959 Johnny Horton recorded and released the number one hit *The Battle of New Orleans*. In the early 21st century, the song got a new lease on life, as a Lego animated video of the song was created by Brickworks and released on YouTube.

Introduction
For Admiral Alexander Cochrane, New Orleans was a prize worth taking. The city's warehouses were filled with goods, which would provide him with a pretty penny in prize money. Its capture would close the Mississippi River from world markets and serve to block further American westward expansion.

The British force was assembled and following the defeat of the American squadron, the way forward was opened. Though the Americans had inflated the number of their troops in the area, the British veterans did not hold the Americans in high esteem. All the British needed was an aggressive or at least competent commander. What they got was Maj. Gen. Sir Edward M. Packenham. Packenham had served in the Peninsula campaign but had not held an independent command. Perhaps his strongest attribute was that he was the Duke of Wellington's brother-in-law and was available. Maj. Gen. Robert Ross was originally chosen to spearhead this expedition, but his death the previous fall precluded his ability to lead the attack.

Packenham did his best to reach his new command, but only arrived after the British had already landed and fought off an American night attack at Villere Plantation. Rather than push on the Americans, Packenham allowed Jackson to select the ground best suited to contain the British assault. At about five miles from New Orleans proper, he began to fortify the Rodriquez Canal – a 1,000-yard millrun between a wood and the Mississippi River. Relying on enslaved labor borrowed from nearby plantations, the Americans had enlarged the canal and used the soil to create breastworks. On the American right flank, they built a redoubt on top of the levee, which would allow any cannon placed their enfilade fire. Along the swampy wooded area, the Americans had built a palisade

151. Gen. Sir Edward Pakenham by Martin Cregan, courtesy of John R. Lee.

152. One simply does not walk into New Orleans - 40mm 95th Rifles, author's photo.

of loopholed logs with a wooden flatform to keep the defender's feet out of the mire. A politically astute general, Jackson named the defensive line The Jackson Line. In the time available, the Americans also fortified the western bank of the river and constructed a second and third defense line, each one closer to New Orleans.

On Christmas day, Packenham arrived on the field and observed his army. His first concern was the presence of two American boats on the Mississippi, which continued to fire on his forces. Though the British heavy artillery had not yet arrived on the field, the British artillery employed heated hot shot and was able to destroy the *USS Carolina*. Lacking horses to outfit the dismounted squadron of Light Dragoons, on 28 December Packenham led a reconnaissance in force to observe the American defenses. The British blundered into the Jackson Line and after suffering casualties from the American land-based artillery as well as from the *USS Louisiana*, Packenham remarked that the defenses were too strong to rush and impractical to flank. Urged by Cochrane and other officers present at Bladensburg, Packenham was assured of the poor quality of American troops and decided to press on with an attack. Packenham then focused on bringing up the British heavy artillery. Bringing up the heavy guns through the swamps was a monumental task, made more difficult as the guns were mounted on naval carriages, rather than on field carriages. During the night of 31 December, the British brought up their guns and constructed artillery firing positions. By early morning of 1 January, the British began their bombardment of The Jackson Line.

The Artillery Battle

The British plan was once the guns had breached the American defenses, a small group of chosen men would provide a firing party and screen Maj. Gen. Samuel Gibb's brigade, which would march in column. Along the British right, a group of 200 men would advance to the edge of the wood and try to draw American troops out from the Jackson line. Once Gibb's men had broken through the line, the remaining British troops would move forward.

The Americans mounted less cannon than their British counterparts but were better situated and had adequate ammunition reserves. The British had limited shells and had no prospect of an immediate resupply. After three hours of firing, the British had exhausted the majority of their ammunition and the American line was none the worse for wear. That evening, the British were able to remove their guns and began to plan for a second assault.

153. 12 pdr. Rocket Troops, courtesy of Pendraken Miniatures.

In the days that followed, more reinforcements appeared. In the British camp, Maj. Gen. John Lambert's brigade arrived, bringing with them additional supplies. On the American side, Maj. Gen. John Thomas arrived with 2,368 men from Kentucky, of which only 700 were armed.

The Battle

Admiral Cochrane suggested that Packenham enlarge the Canal Villere and extend it through the levee to reach the Mississippi. With such a path, sizable British boats from the naval squadron could reach the river from Borne Lake and ferry men to the western side of the river. From this position, the British could take out the American Artillery that faced the British and could turn the captured pieces on their former owners. Once this attack had succeeded, Packenham would attack the main line on the eastern side of the river. The attack would be launched at dawn, when the morning fog would reduce visibility and help hide the advancing British columns.

By 7 January Col. Thornton was given command of the strike force and he had collected the required number of boats in the canal. That evening, as the men pushed their boats forward, the canal collapsed in places, which delayed their progress. Packenham observed that only part of Thornton's force was ready to proceed and redirected those troops not yet ready to launch. The strong river current carried the boats further downstream and Thornton's force landed a considerable distance from the Americans. British deserters had warned Jackson of Packenham's attack, but Jackson was unaware of the attack on the west bank. The area was poorly defended and the enslaved laborers had only partially completed construction of an earthwork.

At 0600, the British launched their attack on the Jackson line. By this time, the Americans were protected by a 4-foot high earthen rampart which was reinforced with wood and bales of cotton. The 44th Foot, which was to lead the assault, was delayed and ended up on the left flank. American fire was devastating on the British officers leading from horseback – Gibbs and Col. Wilkinson both went down. Witnessing the carnage, Packenham shifted the 93rd Highlanders to the left, which stripped Lt. Col. Robert Renny's force of the manpower needed to exploit any breakthrough. Renny stormed a redoubt, but with no troops to exploit his success, became the center of the American fire and died on the spot. Trying to restore order to the advance, Packenham tried to rally the 2nd column but

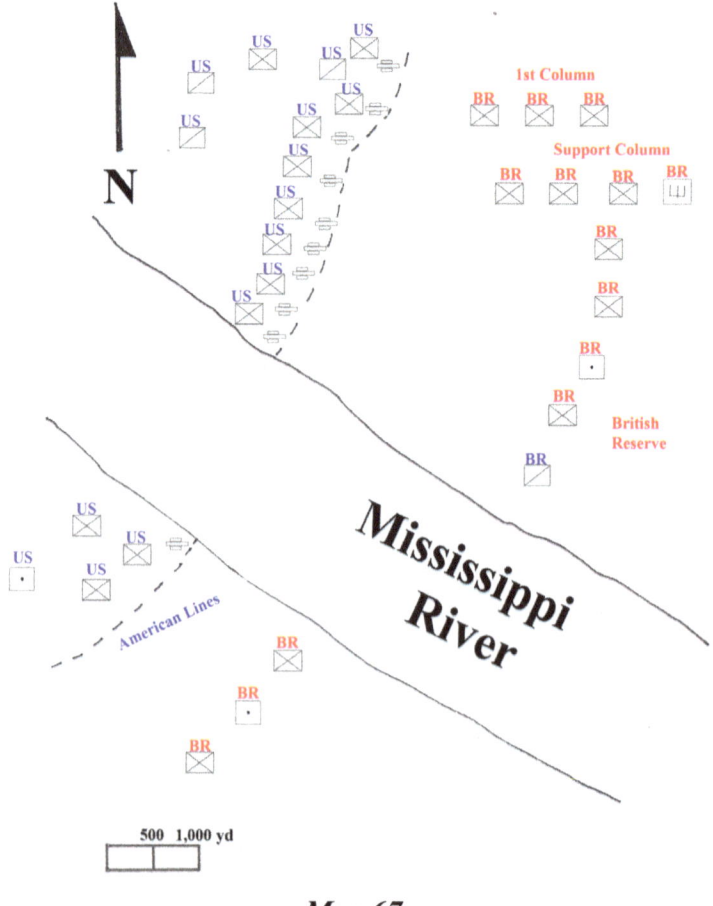

Map 67

made an imposing target. Both Packenham and his horse were brought down.

With Packenham out of the fight, command passed to Maj. Gen. Lambert. Rather than sending more men forward, Lambert took up a makeshift defensive position and awaited an American counterattack. It was then that Thornton was finally able to launch his attack on the west bank.

On the west bank, the American defenses consisted of a few hundred militiamen, three cannons, and partially completed defenses. Thornton's force was able to push through the initial American defenses and moved to carry the secondary line. Though wounded, Thornton was able to move about a mile past the Jackson line and was in a position to threaten New Orleans. Lambert, however, was concerned that the British would be unable to protect themselves from an American counterattack and ordered Thornton to retreat. Thornton's force spiked the captured American iron guns but were able to bring back three small brass cannons, include one that was inscribed as having been captured at Yorktown.

The next day, under a white flag, both sides buried their dead. Lambert was able to remove his troops from the area without Jackson's interference. The one recorded American raid on the British camp apparently was a privately organized affair undertaken to recover the formerly enslaved to whom the British granted freedom and planned to take with them when they left North America.

Order of Battle
British - 10,000 men
Commanding Officer: Maj. Gen. Sir Edward M. Packenham

On the East Bank
1st Column, 2,400 men
Maj. Gen. Sir Samuel Gibbs
44th Foot, 816 men
21st Fusiliers, 790 men
4th Foot, 796 men

Column of Support, 2,650 men
Maj. Gen. Sir John Keane
93rd Foot, 1,100 men
7th Fusiliers, 780 men
43rd Light Infantry, 862 men

Column of Reserve, 5,000 men
Maj. Gen. John Lambert
95th Rifles, 500 men
85th Light Infantry, 560 men
89th Foot, 390 men
Royal Marines, 600 men
Royal Artillery, 318 men
    Two gun batteries
    One Rocket battery
1st West India Regiment, 900 men
14th Light Dragoons, 210 men
Royal Engineers, 60 men

On the West Bank
Col. Sir William Thornton
85th Foot, 300 men
Royal Marines, 200 men

154. Pirate Gun, courtesy of Sash and Saber Castings.

Royal Navy, 200 men

5th West India Regiment, 796 men, redeployed by Packenham to the Column of Reserve
Royal Artillery, redeployed by Packenham to the Column of Reserve
    two 9-pdrs.
    two howitzers

US - 5,400 men
Commanding Officer: Maj. Gen. Andrew Jackson

At the Jackson Line
Maj. Gen. Andrew Jackson
Col. George T. Ross, commanding the American right
The troop placement is from the river, which forms the American right.

New Orleans Volunteer Rifle Company, 36 men
Battery 1 (atop of the old kiln closest, to the river) - two 12pdr howitzers, one mortar
Seventh US Infantry Regiment, 430 men
Battery 2 - one 24-pdr.
Planche's Volunteer Uniform Battalion, 300 men
Battery 3 - two 24-pdrs.
Lacoste's Battalion of Louisiana Men of Color, 280 men
Battery 4 - one 32-pdr.
Daquin's Battalion of Santo Domingo Men of Color, 180 men
Battery 5 - two 6-pdrs.
Forty-Fourth US Infantry Regiment, 350 men

Maj. Gen. William Carroll, commanding the American center
Battery 6 - one 18-pdr. and one 4-pdr.
US Marines, 58 men
Tennessee Volunteers, 800 men

Battery 7 - one 12-pdr. and one 4-pdr.
Kentucky Volunteers, 700 men
Battery 8 - one 9in. howitzer

Brig. Gen. John Coffee, commanding the American far left
Tennessee Mounted Volunteers, 500 men
Choctaws, 62 men

Reserves
Mississippi Cavalry, 150 men
Attakapas Dragoons, 50 men
Horse Volunteers, 1 company, 30 men
Harrison's Battalion, Kentucky militia, 300 men

On the West Bank
Gen. David B. Morgan
1st Defensive Line
Kentucky Volunteers, 200 men

Sixth Louisiana Militia, 100 men

155. US Defensive Line, author's photo.

2nd Defensive Line
Louisiana militia, 150 men
one 12-pdr.
two 6-pdrs.

US Navy West Bank Batteries (facing towards the east bank)
one 24-pdr.
two 18-pdrs.

Game Notes

Cochrane's plan to attack the west bank was brilliant, but too few men and supplies were allocated in time to take full advantage of the situation. Packenham was clearly not up to the task and his micro-management of the battle once it started did little to help. Jackson may have made a number of mistakes, but as he is on record as having won the battle, he became a war hero in the US and for his actions, was latter rewarded with the Presidency.

Given the battle's notoriety among American wargamers, it is often recreated on the tabletop. Unfortunately, in most of the games the author took part in, the British have no way of winning even as they outnumber the US by almost two to one odds. In one game the author and two other players formed up their British troops into a massive assault column and attacked the Jackson line. Having broken through the Jackson line at one point, we were able to turn the American flank and roll up the American defense, which irritated the game master to no end. Not only were the British able to carry the position in three or four turns, but the game which was to run for a three of four hours was completed in less than one. In other historical tabletop recreations of this battle that the author took part in, the Americans won easily, as they simply shot the British to pieces.

Perhaps the most balanced, and thus gameable, aspect of the battle is to refight the western side of the river. While Thornton was able to crush the American defenders with the limited number of troops he had on hand, if he was properly reinforced and the main assault had waited until later in the day, it remains unclear if the Americans would have been able to hold the line as easily as they did.

In this tabletop recreation, Packenham's assault will launch after Thornton's men have attacked the Americans. Thornton's attack can be played as either an independent action or part of the larger battle. Conversely, a hypothetical what-if scenario can be played after Thornton's men have driven off the American defenders and have turned the captured heavy navy cannons of the American defenders. This would result in the American guns closer to the river would either be engaged in an artillery duel and not be able to concentrate their fire on the advancing British or be pummeled without returning fire.

The British brought scaling ladders and fascines to cross the canal ditch, but the delay in the 44th bringing them up resulted in the earthen ditch being a more formidable obstacle than it needed to be. If playing the what if scenario, the 44th would have been in their designated place with the correct gear and any unit would need to spend only one turn crossing the canal. If one were to begin the game following Packenham's timetable, a die roll would be required to determine how many turns would the attacking unit be stuck in the ditch. A die roll of a 1 or a 2 would require a unit to spend 1 turn, a die roll of a 3, 4, or 5 would trap the unit for 2 turns, and a die roll of a 6 on a six would require the unit to be stuck in the ditch for three turns.

The Americans, once set up, mounted a static defense. As such, the game lends itself to solo play. After setting up the Americans, the single player can run the British side and try to breach the Jackson line. Indeed, it may be more of a challenge for all players to take on the role of the British and have the Americans fire automatically when a British unit gets within range. Once a British unit breaches the line, the two nearest American units would try to recapture the ground by either firing or charging. A simple die roll could determine what each unit does when face to face with the enemy – on a roll of a 1,2, or 3, the American unit fires at the enemy, while on a 4, 5, or 6, the unit tries to charge. If a unit loses a melee, the next American unit down the line would become active and would try to force the British unit from the line.

Victory Conditions

This would depend on the scenario. If one where playing the British attack on the river's west side, the British would need to drive the Americans off of the field in twelve turns. The Americans would win if they held the British for that long.

If one were playing the complete game, the victory conditions would be the same - British would win if they drove the Americans from the table – but the game would need to run for twenty turns. If one wanted just to recreate the main attack on the Jackson line, then sixteen turns should be an adequate amount of time to determine a winner.

Further Reading

The Expedition Against New Orleans, in North America. 1814-15. *Journal of the Society for Army Historical Research* 8, no. 32 (1929): 86-113. 1929

Elting, John R., *Amateurs, To Arms!: A Military History of the War of 1812.* (Da Capo Press. New York. 1995)

Pickles, Tim, *New Orleans 1815: Andrew Jackson Crushes the British.* (Osprey Publishing. Oxford. 1993)

Reilly, Robin, *The British at the Gates: The New Orleans Campaign.* (G. P. Putnam's Sons. New York. 1974)

156. British Foot attacking a Blockhouse, authors photo.

Appendix 1

Available Figures

War of 1812 miniatures - An incomplete list –

Not surprisingly, almost all manufacturers produce British line that works for the War of 1812. Though a number of companies produce figures for the American line and US rifles, these are harder to find and many gamers represent American militia with figures from other ranges, including the American War of Independence and the Alamo. A list of wargame manufactures is presented below.

For those who which to create their own figures, it is quite easy to create the 1813 American Infantry troops using British Waterloo figures. This is especially true when using smaller scale figures. Difference between the 1813 US uniform, which was worn for most of the conflict, and the uniforms worn by British troops at Waterloo are minimal, with the greatest difference being in the backpacks. The British pack had the blanket rolled on top while the American pack allows for the blanket to be folded underneath the outer flap. An article in *Wargames Illustrated* No. 260 discussed how to convert 28mm British plastic figures to American troops that fought in Maryland.

In smaller scales, such as 10mm, it is very easy to convert a British figure to an American infantryman. Simply clip the rolled blanket from the top of the figure and paint in appropriate colors. The photograph below of 10mm Pendraken figures shows the converted figure and how they look when painted as US Infantry.

54mm

A Call To Arms

Older British company that made British Guards, Lights, the 95th, and artillery for Waterloo.

Airfix

Ancient semi-hard plastic figures of British elites. Have been recast over the last few decades by various companies.

Armies In Plastic

Reasonably priced US and British figures. Produced US Infantry, British Foot, Native Americans and artillerymen. Plastic Native Americans were sculpted by John Jenkin sculpts and definitely worth picking up. Unfortunately, the US Line and British Foot are very basic sculpts that are sold under different names depending on the color of the plastic.

Barzo

Long ago, Barzo made a War of 1812 playset that included cotton walls and a Congreve Rocket battery. LOG is currently re-releasing Barzo sets and may produce the figures again. The original figures were brittle.

Expeditionary Force

New guys on the market. Limited releases of US line, Kentucky volunteers, British line and Royal Marines. Well detailed and good quality semi-hard plastic figures. No guys carrying flags and rather expensive at $35 for 9 figures.

Italeri

Currently out of production semi-hard plastic figures, but they made 95th Rifles, British Dragoons, and Mounted British Officers.

Victrix

One box of British elites. These are larger versions of their Heroic Scale 28mm line. At one time they also sold British flag bearers, but these have recently disappeared from their web site. Hard plastic.

40mm

Perry Miniatures
Have a metal range for the Peninsula. The British work well for the War of 1812.

Sash and Saber
Probably one of the most complete ranges of War of 1812 figures on the market. Only noticeable gaps are the lack of Naval landing parties, African-American heads and Congreve Rocket batteries.

Trident Miniatures
Made a variety of US militia and Naval landing figures within their extensive American War of Independence range. Recently sold so they may be available again on the market.

25/28mm

Brigade Games
They produce a variety of American and British figures. They appear to work well Knuckleduster.

Dixon
Trevor Dixon is best known for his specific sculpting style and erratic releases. The handful of 25mm British figures they released for their Napoleon in Egypt range work as members of the Royal Veterans Regiment.

Front Rank
The make some wonder 28mm British figures. The range is currently for sale, so its unclear if they will be around in the future.

Knuckleduster
They produce 28mm Americans, British and Canadian troops specifically for the War of 1812. The British in Overcoats, Marines, and US Rifles are very useful figures. They also produce Regimental Economy packs.

Minifig
These 25mm figures are now considered "retro" and the definition of old school wargaming. Their British releases dominated wargaming in the 1980/90s and still can be found in people's armies.

North Star
They have a few War of 1812 figures for their Muskets and Tomahawk range. The pictures look nice on the web site, but I am unsure how they work with other ranges.

Old Glory
At one time Old Glory was the poor man's alternative to Wargames Foundry. Economical, at least in the US, they often appear on tabletops. Unfortunately, they run the molds past their prime and depending on the when your figures were cast, the cleanup can take as long as painting the figures. The figures also have a unique dramatic style, which recall the old Hinchcliffe/Elite figures.

Perry Miniatures
The prestigious Perry brothers have definitely made a mark on the wargaming world of today. Though they do not make War of 1812 figures specifically, they make British troops for both Waterloo and Egypt. Pictures of their landing craft are truly awe inspiring, though some non-believers criticize their figures for having little heads.

Redoubt
Not sure if they are still in production. Their larger figures are now dated, but having painted two of their landing boats, they looked very imposing. Figures are big and bulky.

Victrix
Part of the plastic revolution. Great British 28mm figures.

Wargames Foundry
Aly Morrison sculpted an extensive range of 25mm War of 1812 figures in the late 1980s/early 1990s for Foundry. Originally sold as individual figures but Foundry transitioned to blisters in the

late 1990s/early 2000s. The range mostly focuses on the US 1812 uniform, which was obsolete by 1813, but included US riflemen in both dress uniforms and fringe jackets. They also made dismounted US dragoons. Like the rest of what Foundry sells, its hit or miss what is currently in production.

Warlord

This figure manufacturer, along with Perry Miniatures, pioneered the shift to 28mm plastic wargaming. They make hard plastic British line and cavalry for both the Peninsula and Waterloo campaigns.

20mm

Airfix

Made British Infantry, Highlanders, artillery, and cavalry in soft plastic for their Waterloo set in the 1970s. Their horses never stayed glued to their base.

EMHAR

Released a box of British line in the Peninsular uniform. The poses were very useful for wargamers.

ESCI

After Airfix stopped production, ESCI made a slightly taller set of British troops for Waterloo.

Hat

Make British Light Infantry, Peninsular British, Royal Marines and sailors.

Revell

Released a set of Waterloo British and 95th rifles. Rumored to be out of production.

Strelets

A Ukrainian company that makes the only War of 1812 specific US troops in this scale. They also make British in overcoats and Tecumseh Native Americans.

15/18mm

AB Figures

Sculpted by Anthony Barton, the Napoleonic range includes a line of 18mm British figures that many gamers truly love. Often referred to as the best high end of the 18mm. Expensive and rarely scene in the US.

Battle Honors

True 15mm figures sculpted earlier by Anthony Barton. Currently being produced in the US by 19th Century Miniatures.

Blue Moon/Old Glory

Sculpted by Chris Hues, the US in the War of 1812 range is rather complete and now is up to eighteen packs. If you become a member of the Old Glory Army, the figures become even more affordable.

19th Century Miniatures

These are older Old Glory figures now manufactured by 19th Century Miniatures. They come in packs of 24 figures with command. They produce both British and US forces.

Minifig

They made everything at one time, including LOTS of Napoleonic British as well as an American range of figures. Currently owned by Caliber Books.

13.5mm Epic Scale

Warlord Games

Produce an extensive range of figures based on the battle of Waterloo in hard plastic. With some modification the British foot are easily converted to US Infantry. (see Appendix 4)

10mm and Smaller

Alder – 6mm
They have an extensive range of British figures, which can be also be painted as US troops.

Baccus – 6mm
They produce an American and British range for the War of 1812. To get you started, they also produce an American Army Pack.

Perry Miniatures – 8mm
The Perry's produce a Travel Battle game. This box set comes with rules, terrain, and hard plastic figures which can be painted as British and Americans.

Pendraken – 10mm
British produced figures that have a great deal of detail. No specifically designed US War of 1812 figures but with a simple snip a British infantryman becomes an American infantryman. Also, their cavalry is on piece, which makes it much easier.

Old Glory – 10mm
The infantry figures come in strips, which makes for easy painting. They produce a range of British for the Peninsula, which at this scale can easily be converted to US infantry. The only downside is that the cavalry. Unfortunately, need to be put together, which can require a good deal of carving to get the saddle to fit on the horse.

Boats

GHQ
Best known for making micro-armor, they carry a range of metal 1:1200 scale Napoleonic ships. They also make a Baltimore Privateer, which is very cool.

Langton Miniatures
Produce stunning models in 1:300 scale and a range of 1:1200 scale boats, including the USS *General Pike*.

Old Glory
15mm range of War of 1812 boat includes a number of vessels, including the USS *Niagara*. They also produce 1:300 scale boats.

Sails of Glory
This award winning is a tactical ship to ship game comes with pre-built and pre-painted 1:1000 scale plastic ships. The range focuses primarily on larger craft but does include one 14 gun sloop.

Valiant Enterprises Ltd.
They made 1:2000 scale ships which fit the old Avalon Hill Wooden Ships and Iron Men boardgame. The line is still being produced, but now in pewter.

Warlord
Released in 2019, Black Sails is a naval game that uses 1:700 plastic boats. They produce both US and British fleets, as well as gun emplacements in their Scenery pack. In addition to the larger boats, they also produce schooners, brigs, sloops, cutters, and gunboats.

Appendix 2 - List of Illustrations

Note - (Lossing) is Benson Lossing's Pictorial Field-book of the War of 1812, which was originally published in 1868

Front Cover - 40mm Sash and Saber Miniatures, author's photo
1. British Light Dragoons, courtesy of Perry Miniatures. pg. 4
2. US Rifle Regiment courtesy of Richard Watts. pg. 5
3. Incorporated Militia, courtesy of Knuckleduster Miniatures pg. 6
4. A View of the Old Church in St. Regis, (Lossing) pg. 7
5. Native Americans, courtesy of Craig Charron. pg. 9
6. New State Militia, courtesy of Richard Watts. pg. 10
7. Red George Mac Donnell, courtesy of John R Lee. pg. 12
8. British in overcoats, courtesy of Roger Chrysler. pg. 14
9. US Militia artillery, courtesy of Richard Watts pg. 15
10. 1814 Drawing of a Bateaux, Mrs. Edward Kemp Collection F360 OS 1-8 Archives of Ontario. pg. 17
11. New York Militia, courtesy of Mike Vasile. pg. 19
12. British Infantry, courtesy of Richard Watt. pg. 20
13. Charles de Salaberry, courtesy of John R. Lee. pg. 21
14. Canadian Fencibles, courtesy of Perry Miniatures. pg. 23
15. US Infantry, author's collection. pg. 24
16. John de Barth Walbach (Ft. Monroe, 1846). pg. 26
17. Canadian Voltigeurs, courtesy of Sash and Saber Casting. pg. 28
18. US Light Dragoons, courtesy of Knuckleduster Miniatures. pg. 29
19. 49th Foot, courtesy of Roger Chrysler. pg. 30
20. La Colle Mill and Blockhouse (Lossing). pg. 31
21. British Rocket Battery, courtesy of Richard Watt. pg. 34
22. US Artillery, courtesy of Knuckleduster Miniatures. pg. 35
23. Thomas MacDonough by Thomas Gimbreder. (Naval History & Heritage Command). p.36
24. Sir George Prevost, courtesy of John R Lee. pg. 40
25. British Squadron, Warlord Miniatures. pg. 42
26. 58th Foot, courtesy of Perry Miniatures. pg. 44
27. Rocket Battery, courtesy of Knuckleduster Miniatures. pg. 44
28. US troops in prepared positions, courtesy of Mike Vasile. pg. 45
29. Zebulon Montgomery Pike by Charles Wilson Peale 1808, National Portrait Gallery. pg. 47
30. Ft. York, C.W. Jeffrey. pg. 48
31. US Infantry on the march, Pendraken conversions, author's photo. pg. 50
32. Royal Newfoundland Fencibles, courtesy of Perry Miniatures. pg. 51
33. 3rd US artilleryman armed with a musket, courtesy of W Britain. pg. 53
34. Author's Sacketts Harbor game, circa 1996. pg. 54
35. US Light Dragoons (dismounted), courtessy of Richard Watts. pg. 55
36. British landing, courtesy of Perry Miniatures. pg. 56
37. Local Militia, courtesy of Tod Kershner pg. 58
38. Royal Marines, author's photo. pg. 59
39. Sailors Skirmishing with Muskets, courtessy of Perry Miniatures. pg. 62
40. New York State Militia Officer, courtesy of W Britain. pg. 63
41. Royal Navy Landing Boat with Crew, courtesy of Britannia Miniatures. pg. 63
42. Fort Ontario, author's photo. pg. 65
43. Ft Oswego 1814 Garrison Flag, Drummond family Castle, Courtesy of Paul Lear 1814. pg. 66
44. De Watteville's Regiment, courtesy of W Britain. pg. 66
45. US Rifles, author photo. pg. 68
46. Oneidas, courtesy of Richard Watts. pg. 69
47. Royal Marine Firing Line, courtesy of Perry Miniatures. pg. 69
48. HMS St Lawrence, courtesy of John R. Lee. pg. 70
49. 1st Rate and Frigates on Lake Ontario, October 1814, courtesy of Chris Ngiau. pg. 72
50. Plan of Ft Henry, drawn by Christian Wright. pg. 73
51. Fulton's Torpedo Boat in action, courtesy of John R Lee. pg.74
52. US Infantry, courtesy of Knuckleduster Miniatures. pg. 75
53. Royal Artillery, Courtesy of Perry Miniatures. pg. 76
54. Capt. Elliot by David Edwin 1813, Naval History and Heritage Command. pg. 78
55. Small boats, author's photo. pg. 80
56. Excelsior Advances, 40mm New York State Militia, author's photo. pg. 81
57. British Infantry, courtesy of Sash and Saber Castings. pg. 81
58. Isaac Brock Action Figure, author's photo. pg. 84
59. View from atop the Niagara Escarpment toward the American Embarkation Area, author's photo. pg. 85
60. US Infantry in Early Uniforms, authors photo. pg. 86
61. Maj. Gen. Winfield Scott circa 1820, National Portrait Gallery. pg. 88
62. American Landing Site, author's photo. pg. 90
63. Ft George, author's photo. pg. 91
64. Canadian Fencibles, courtesy of Richard Watts. pg. 92
65. Stoney Creek Battlefield circa 1860 (Lossing). pg. 93
66. US Artillery, courtesy of Roger Chrysler. pg. 95
67. Fog of War, courtesy of Mike Vasile. pg. 96
68. Laura Secord C.W. Jeffrey. pg. 98
69. Native Americans, courtesy of Richard Watts. pg. 100
70. Fourteenth Infantry being flanked by Native Americans, courtesy of Roger Chrysler. pg. 101
71. Plan of Ft Schlosser, drawn by Christian Wright. pg. 104
72. New York State Militia, courtesy of Richard Watts. pg. 104
73. Gen. Peter Porter, (Lossing). pg. 106
74. American militia and Native allies, author's photo. pg. 107
75. 8th Foot, courtesy of Knuckleduster Miniatures. pg. 107
76. Commodore Isaac Chauncey, USN by Gilbert Stuart circa 1818 US Naval Academy Museum Collection. pg. 109
77. USS General Pike by Charles Ware circa 1820, National Archives. pg. 110
78. Sir Phineas Riall, courtesy of John R. Lee. pg. 112
79. Scott's Brigade, courtesy of Knuckleduster Miniatures. pg. 114
80. 100th Infantry Grenadier Company, author's photo. pg. 115
81. US Light Dragoons, courtesy of Knuckleduster Miniatures.pg. 115
82. Jacob Brown by John Wesley Jarvis circa 1815, National Portrait Gallery. pg. 116
83. Lundy's Lane Battlefield, author's photo. pg. 117
84. 19th Light Dragoons, courtesy of Perry Miniatures. pg. 118

85. 11th US Infantry, courtesy of Knuckleduster Miniatures. pg. 119
86. British Foot Artillery, courtesy of Perry Miniatures. pg. 120
87. Gen. Sir Gordon Drummond by George Theodore Berthon 1883, Government of Ontario Art Collection. pg. 121
88. Ft Erie, author's photograph. pg. 123
89. New York State Militia wearing their distinctive red headscarves, author's photo. pg. 125
90. US Infantry, author's photo. pg. 126
91. 1st Royal Scotts, courtesy of Knuckleduster. pg. 128
92. Daniel Bissell, courtesy John R. Lee. pg. 131
93. Cook's Mill Monument, courtesy John R. Lee. pg. 131
94. Glengarry Light Infantry, courtesy of Tod Kershner. pg. 132
95. US Rifle Regiment, courtesy of Roger Chrysler. pg. 132
96. Frenchtown Battlefield circa 1860 (Lossing). pg. 134
97. Sled gun, courtesy of Ralph Naveaux. pg. 136
98. Native Americans, courtesy of Tod Kershner. pg. 137
99. Kentucky Volunteer Rifles, Command, courtesy of Knuckleduster Miniatures. pg. 137
100. William Henry Harrison by Rembrandt Peale circa 1813, National Portrait Gallery. pg. 138
101. Ft Meigs, author's photo. pg. 140
102. Native Americans, courtesy of Roger Chrysler. pg. 141
103. Outside the walls of Ft. Meigs, courtesy of Harold Oney. pg. 143
104. Frontier Militia in Hunting Shirts, courtesy of Sash and Saber Castings. pg. 145
105. Oliver Hazard Perry by Jane Stuart circa 1857, Birmingham Museum of Art. pg. 146
106. Battle of Lake Erie, C. W. Jefferys. pg. 148
107. The restored U.S. Brig Niagara demonstrates 1813 era naval gunnery, photo by John Baker, courtesy of the Erie Maritime Museum. pg. 149
108. We have met the enemy and they are ours ..., courtesy of Chris Ngiau. pg. 149
109. Battle of the Thames, C. W. Jefferys. pg. 150
110. Recaptured 3-pdr cannon, author's photo. pg. 152
111. Kentucky Mounted Infantry, courtesy of Knuckleduster Miniatures. pg. 152
112. British Line, courtesy of Pendraken. pg. 153
113. Ft Mackinac (Lossing). pg. 154
114. 40mm US Infantry, author's photo. pg. 155
115. British troops, courtesy of Tod Kershner. pg. 156
116. British Landing, courtesy of Perry Miniatures. pg. 160
117. Militia in Civilian Dress, courtesy of Sash and Saber Castings. pg. 161
118. Royal Marines, courtesy of Richard Watts. pg. 161
119. Rear-Admiral Sir George Cockburn by John James Halls circa 1817, National Maritime Museum. pg. 163
120. Battlefield monument at Havre De Grace, author's photo. pg. 164
121. Sailors with pistols, courtesy of Perry Miniatures. pg. 164
122. British Flat boat, courtesy of Perry Miniatures. pg. 166
123. USS Constellation, courtesy of GHQ Models. pg. 167
124. Joshua Barney, (Lossing). pg. 168
125. Bladensburg, view from the British position author's photo. pg 170
126. British Light Troops skirmishing, courtesy of Perry Miniatures. pg. 171
127. US Marines, courtesy of Richard Watts. pg. 171

128. Royal Artillery Rocket Brigade, courtesy of Knuckleduster Miniatures. pg. 172
129. Peter Parker, (Lossing). pg 173
130. Sailors attacking with boarding pikes, courtesy of Perry Miniatures. pg. 174
131. Cpt. Peter Parker and Royal Marines, courtesy of Richard Watts. pg. 176
132. Gen. Robert Ross, courtesy of John R Lee. pg. 177
133. First Baltimore Sharpshooters, courtesy of Richard Watts. pg. 158
134. Fifth Maryland, courtesy of Richard Watts. pg. 182
135. 15mm British Infantry, courtesy of Dave Abraham. pg. 183
136. Royal Rocket Artillery, courtesy of Warlord Games. pg. 184
137. Royal Navy Landing Party, courtesy of Wargames Foundry. pg. 186
138. Virginia militia, authors photo. pg. 186
139. View toward Point Peter and St Mary's from Cumberland Island, courtesy of Cumberland Island National Seashore, National Park Service. pg. 187
140. US Rifle Regiment, courtesy of Roger Chrysler. pg. 189
141. 2nd West Indian Regiment, courtesy of Tod Kershner. pg. 189
142. Plain of Ft Bowyer, drawn by Christian Wright. pg. 192
143. Native Americans courtesy of Mike Vasile. pg. 193
144. US Infantry wearing white linen or cotton jackets, author's photo. pg. 194
145. US Gunboats of the types used during the War of 1812, 19th century print, Naval History and Heritage Command. pg 196
146. Gunboat Squadron, courtesy of Warlord Games. pg. 197
147. A swarm of gunboat, author's photo. pg 197
148. Andrew Jackson by Ralph Eleaser circa 1817, National Portrait Gallery. pg. 198
149. 95th Rifles, author's photo. pg. 200
150. US Infantry, courtesy of Knuckleduster Miniatures. pg. 201
151. Gen. Sir Edward Pakenham by Martin Cregan, courtesy of John R. Lee. pg 202
152. One simply does not walk into New Orleans - 40mm 95th Rifles, author's photo. pg. 203
153. 12 pdr. Rocket Troops, courtesy of Pendraken Miniatures. pg. 204
154. Pirate Gun, courtesy of Sash and Saber Castings. pg 205
155. US Defensive Line, author's photo. pg. 207
156. British Foot attacking a blockhouse, author's photo, pg. 209
157. 1800 pattern British shako plate, courtesy of Craig Charron. pg. 217
158. 1812 pattern British shako plate, courtesy of Craig Charron. pg. 217
159. Original British casting. pg. 218
160. Pendraken 10mm British Infantryman with clipped backpack, author's photo. pg. 218
161. US Infantryman, author's photo. pg. 218
162. Painted British and American Infantrymen, author's photo. pg. 218
163. Old Glory ships, author's photo. pg. 219
164. Native Americans, courtesy of Tod Kershner. pg. 219
165. US Navy and Marines, courtesy of W Britain. pg. 219
166. US Infantry, courtesy of Knuckleduster Miniatures. pg. 219
Back Cover - British Indian Pattern "Brown Bess" musket, author's collection

Appendix 3 - List of Maps

Map 1 - St. Lawrence/ Eastern Lake Ontario, pg. 3
Map 2 - Raid on Gananoque, pg. 4
Map 3 - St Regis, pg. 8
Map 4 – Ogdensburg, pg. 13
Map 5 - Cranberry Creek, pg. 18
Map 6 - Battle of Chateauguay, pg. 22
Map 7 - Crysler's Farm, pg. 25
Map 8 – Crysler's Farm, pg. 28
Map 9 - Lacolle Mill, pg. 32
Map 10 – Plattsburgh, pg. 37
Map 11 – Plattsburgh, pg. 38
Map 12 – Plattsburgh, pg. 39
Map 13 – Plattsburgh, pg. 41
Map 14 – Plattsburgh, pg. 43
Map 15 - Central Lake Ontario, pg. 46
Map 16 – York, pg. 49
Map 17 - Sackett's Harbor, pg. 52
Map 18 - Sodus Point, pg. 57
Map 19 – Pultneyville, pg. 61
Map 20 – Oswego, pg. 64
Map 21 - Big Sandy, pg. 67
Map 22 – Kingston, pg. 70
Map 23 – Kingston, pg. 74
Map 24 - Niagara Peninsula/ Western Lake Ontario, pg. 77
Map 25 - HMS Caledonia and HMS Detroit, pg. 79
Map 26 - Queenston Heights, pg. 83
Map 27 - Queenston Heights, pg. 87
Map 28 - Ft. George, pg. 89
Map 29 - Stoney Creek, pg. 94
Map 30 - Beaver Dams, pg. 99
Map 31 - Raid of Ft. Schlosser, pg. 103
Map 32 – Black Rock, pg. 105
Map 33 - Burlington Races, pg. 108
Map 34 – Chippawa, pg. 113
Map 35 - Lundy's Lane, pg. 117
Map 36 - Ft. Erie, pg. 122
Map 37 – Ft. Erie, pg. 124
Map 38 – Ft. Erie, pg. 126
Map 39 – Ft. Erie, pg. 127
Map 40 - Cook's Mill, pg. 130
Map 41 – Western Great Lakes, pg. 133
Map 42 - The River Raisin, pg. 135
Map 43 - Ft. Meigs, pg. 139
Map 44 - Ft. Meigs, pg. 142
Map 45 - Ft. Meigs, pg. 142
Map 46 - Ft. Meigs, pg. 144
Map 47 - Ft. Meigs, pg. 145
Map 48 - Put-In-Bay, pg. 147
Map 49 – Thames, pg. 151
Map 50 - Mackinac Island, pg. 155
Map 51 – East Coast, pg. 158
Map 52 - Raid on French Town, pg. 159
Map 53 - Havre de Grace, pg. 162
Map 54 - Craney Island, pg. 165
Map 55 – Bladensburg, pg. 169
Map 56 - Caulk's Field, pg. 174
Map 57 - North Point, pg. 178

Map 58 - North Point, pg. 179
Map 59 – North Point, pg. 181
Map 60 – North Point, pg. 182
Map 61 - Kirby's Windmill, pg. 185
Map 62 - Point Peter, pg. 188
Map 63 – Gulf Coast, pg. 191
Map 64 - Ft. Bowyer, pg. 193
Map 65 - Battle of Lake Borgne, pg. 195
Map 66 - Villere Plantation, pg. 199
Map 67 - New Orleans, pg. 204

157. 1800 pattern British shako plate, courtesy of Craig Charron.

158. 1812 pattern British shako plate, courtesy of Craig Charron.

Appendix #4
Converting Pendraken 10mm British Foot to US Infantry

159. Original British casting.

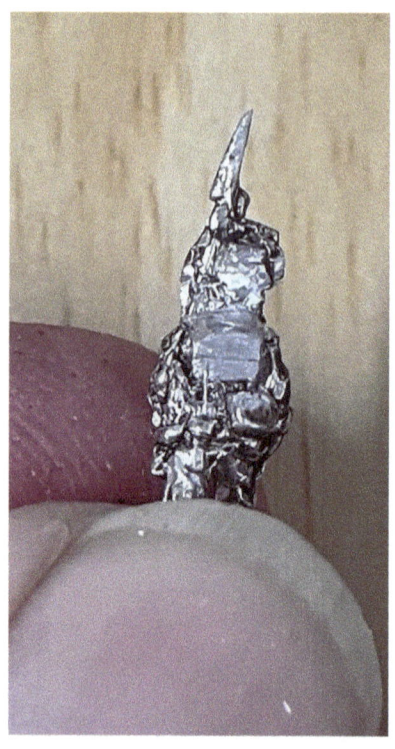

160 Pendraken 10mm British Infantryman with clipped backpack, author's photo.

161 US Infantryman, author's photo.

162 Painted British and American Infantrymen, author's photo.

163. Old Glory ships, author's photo.

164. Native Americans, courtesy of Tod Kershner.

165. US Navy and Marines, courtesy of W Britain.

166. US Infantry, courtesy of Knuckleduster Miniatures.

Look for more books from Winged Hussar Publishing, LLC – E-books, paperbacks and Limited-Edition hardcovers. The best in history, science fiction and fantasy at:

https://www.whpsupplyroom.com

or follow us on Facebook at:

Winged Hussar Publishing LLC

Or on twitter at:

WingHusPubLLC

For information and upcoming publications

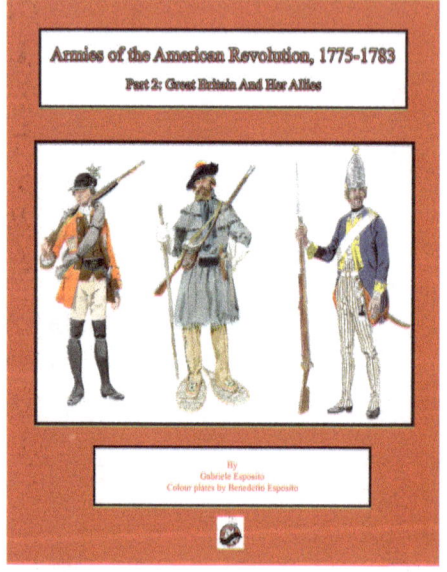

www.ingramcontent.com/pod-product-compliance
Lightning Source LLC
Chambersburg PA
CBHW061124010526
44114CB00029B/3001